Myth-Busting
in Science and Religion

Myth-Busting in Science and Religion

An Introduction for Ignorant Intellectuals

Jim Slagle

 CASCADE *Books* • Eugene, Oregon

MYTH-BUSTING IN SCIENCE AND RELIGION
An Introduction for Ignorant Intellectuals

Copyright © 2025 Jim Slagle. All rights reserved. Except for brief quotations in critical publications or reviews, no part of this book may be reproduced in any manner without prior written permission from the publisher. Write: Permissions, Wipf and Stock Publishers, 199 W. 8th Ave., Suite 3, Eugene, OR 97401.

Cascade Books
An Imprint of Wipf and Stock Publishers
199 W. 8th Ave., Suite 3
Eugene, OR 97401

www.wipfandstock.com

PAPERBACK ISBN: 979-8-3852-3614-5
HARDCOVER ISBN: 979-8-3852-3615-2
EBOOK ISBN: 979-8-3852-3616-9

Cataloguing-in-Publication data:

Names: Slagle, Jim [author].

Title: Myth-busting in science and religion : an introduction for ignorant intellectuals / by Jim Slagle.

Description: Eugene, OR: Cascade Books, 2025 | Includes bibliographical references and index.

Identifiers: ISBN 979-8-3852-3614-5 (paperback) | ISBN 979-8-3852-3615-2 (hardcover) | ISBN 979-8-3852-3616-9 (ebook)

Subjects: LCSH: Religion and science—History. | Religion and science. | Christianity. | Bible and science.

Classification: BL240.3 S53 2025 (paperback) | BL240.3 (ebook)

VERSION NUMBER 07/10/25

All Scripture quotations, unless otherwise indicated, are taken from the Holy Bible, New International Version®, NIV®. Copyright © 2011 by Biblica, Inc.™ Used by permission of Zondervan. All rights reserved worldwide. www.zondervan.com. The "NIV" and "New International Version" are trademarks registered in the United States Patent and Trademark Office by Biblica, Inc.™

Scripture quotations taken from the (NASB®) New American Standard Bible® are copyright © 2020 by The Lockman Foundation. Used by permission. All rights reserved. lockman.org.

The World's Last Night by C. S. Lewis copyright © 1960 C. S. Lewis Pte. Ltd. 1950. *The Problem of Pain* by C. S. Lewis copyright ©1940 C. S. Lewis Pte. Ltd. 1950. Extracts reprinted by permission.

For Krista, Joah, Kira, and Riki

It is ambition enough to be employed as an under-labourer in clearing the ground a little, and removing some of the rubbish that lies in the way to knowledge.

—John Locke

Contents

Preface | ix

Chapter 1: The Boring Introductory Chapter | 1
Chapter 2: The Shape of Things to Come | 17
Chapter 3: A Central Issue | 37
Chapter 4: Cosmos of Unusual Size | 53
Chapter 5: Medical History | 70
Chapter 6: Tales of the Righteous Dead | 83
Chapter 7: Galileo Figaro Magnifico | 98
Chapter 8: Evolutionary Myths | 112
Chapter 9: The God Killer | 128
Chapter 10: Putting People in Their Place | 149
Chapter 11: Xenomorphobia | 161
Chapter 12: A Brief History of Science and Religion | 178
Chapter 13: Miracle Panegyrical | 194
Chapter 14: Theism and the Justification of Science | 212

Conclusions | 227
Appendix: The Christ Myth Myth | 235
Bibliography | 257
Index | 273

Preface

I. SCIENCE AND TIGERS AND BEARS

THROUGHOUT INTELLECTUAL HISTORY, MANY of the most brilliant people have believed in God. In fact, this was the majority position among the intelligentsia until relatively recently. Yet contemporary society often confronts us with the claim that belief in God is not merely incorrect; it is *obviously* incorrect—it is silly, foolish, inane. This is a strong claim, not least because it would entail that the great thinkers of previous ages were not merely misguided but dumb, and that our knee-jerk reactions are a more reliable guide to truth than their lifelong reflections. But so what? The idea that there might be an omnipotent, omniscient, omnibearded old man in the sky can be dismissed out of hand.

Apart from the beard, however, this is difficult to accept on its face. How could such profound thinkers be so spectacularly mistaken while we can just dismiss their ideas without expending any intellectual energy? To make this claim more plausible, more palatable, we must suggest that their error was not one of reflection but of information: we have knowledge that they, through no fault of their own, lacked. Thus, their belief in God is understandable. Given their evidence base, it made sense to believe in God. However, *our* evidence base has expanded upon theirs to an almost unimaginable extent. It is this additional evidence to which we have access and they did not that makes belief in God no longer plausible, and actually absurd.

What is this new information we have that they lacked? Has there been some new argument that has successfully demonstrated the non-existence of God? Was there some philosophical breakthrough that rendered all prior philosophizing worthless? I suppose that's *possible*, but it would be difficult to claim that such a breakthrough in speculative reasoning became so widely accepted as to produce the effect in question. A philosophical argument, after all, requires the engaging of the intellect, which is pretty much the opposite of a knee-jerk reaction. Not to mention that the primary arguments against God are the same today as they were then: the occurrence of evil and the apparent absence of God (divine hiddenness).[1] People of earlier ages were as familiar with these issues as we are, if not more so. They lived in a world where it was at least as likely that one would experience horrific evils as it is in ours. And if God does not exist, his absence would be felt just as much in earlier eras as our own. There's nothing new here.

No, when we ask what new information we have that the great intellects of the past lacked, *must* have lacked, that explains their otherwise inexplicable belief in God, the answer is not in some new philosophical argument or some appreciation of evil that they failed to see. The answer, in a word, is *science*. We have made scientific discoveries that explain the same things God was alleged to explain but do so in a much better way. Moreover, these scientific discoveries are not matters of speculative philosophy but empirical discovery. One can still deny them, refusing to look through Galileo's telescope as it were,[2] but the evidence is publicly available for those willing to see it. Science has exorcized God from the world.

Of course, the battle is not yet won. There are still many forces in society that fear science, that reject science, or perhaps more contemptibly, accept science but reject the charge that science has somehow banished religion. But this charge is so evident, so blindingly obvious, that argument is superfluous. For some, at least, all that is necessary is to shout "Science! Reason! So there!" as loudly and as often as possible. Because of this attitude, the arguments for the existence of God, religious thought in general, and contemporary attempts to reconcile science and religion need not be given a second thought. Scientific discoveries have provided empirical proof that religious arguments, however clever they may be, are

---

1. The problem of evil is a huge subject. For a survey, see Peterson, *Problem of Evil*. On divine hiddenness, see Howard-Snyder and Green, "Hiddenness of God."

2. This refers to the common accusation that the pope refused to look through Galileo's telescope to observe his discoveries because they challenged the Catholic church's theological claims. In reality, it wasn't the pope, it was two Aristotelian philosophers and scientists, and it's not clear why they refused. One of them, Cesare Cremonini, said when he'd looked through other telescopes, they had given him headaches (Hannam, *God's Philosophers*, 312). The fact that it conflicted with *Aristotle*—not Christianity—didn't have anything to do with it, at least so he said.

spurious. They *must* be spurious since they establish conclusions at odds with science.

Naturally, most people who reject religion because of science are not thoughtless and obnoxious about it. Part of the problem here (as well as every other area of society) is that the most ill-behaved people making the most extreme claims are the ones who get the most attention, since such behavior and claims are more newsworthy, more *interesting*. While I don't want to treat such people as representative, I'm afraid there are such people, and they are legion. Plus, the polite people who agree with the claims in question will still be addressed in what follows.

At any rate, left aside in all this is the real possibility that this Conflict Thesis has been greatly exaggerated. Perhaps science and religion do not conflict. Or perhaps the conflicts are superficial: matters that should not matter. Perhaps it is not science but *scientism*, the view that science is the one and only linchpin to reality, that conflicts with religion. Or perhaps science and religion can offer mutual *support* for each other. Crazy talk, I know.

II. OUTLINE OF THE PRESENT WORK

This book contests the claims of serious conflict between science and religion and seeks to encourage educated laypeople to reexamine their relationship, as well as reopening the door to the serious consideration of traditional reasoning, both pro and con, about God. This is not meant as a defense but as a clearing away of a major stumbling block to giving these ideas their fair shake. Thus, this book is a collection of *prolegomena*, not an *apologia*. "Prolegomena" refers to the stuff we have to study before we start to study other stuff. In none of this will I dispute any scientific theory or evidence, although some popular interpretations of them (*philosophical* interpretations, not scientific) may be challenged.

The subtitle of this book is *An Introduction for Ignorant Intellectuals*. The reason for this is that intellectuals often mistakenly think they know what they're talking about regarding science and religion; so, in this sense, I'm carrying on Socrates's ministry of explaining to people that they don't know what they think they know.[3] And I single out intellectuals because they are naturally prone to do this. They (we) unfortunately

---

3. See, e.g., Plato's *Apology* 21a–23c.

have a reputation. "An intellectual is a person knowledgeable in one field who speaks out only in others."[4]

> Many public intellectuals have been justly renowned within their respective fields but the point here is that many *did not stay within their respective fields*. . . . The fatal misstep of such intellectuals is assuming that superior ability within a particular realm can be generalized as superior wisdom or morality over all. Chess grandmasters, musical prodigies and others who are as remarkable within their respective specialties as intellectuals are within theirs, seldom make that mistake. . . . What many intellectuals seem not to understand is that even being the world's leading authority on a particular subject, such as admiralty law or Mayan civilization, does not confer even minimal competence on other subjects, such as antitrust law, environmental issues, or foreign policy.[5]

There's a type of fallacy called appealing to authority. What makes it a fallacy is not the appeal per se but that those being appealed to are not authorities in the subject in question. They may be authorities in something *else*, but that just means their authority is irrelevant to the subject under discussion.[6] Plenty of intellectuals are guilty of a sort of reflexive appeal to authority—they consider themselves to be authorities in subjects of which they are not. Of course, they're just as free to have strong opinions on issues as anyone else, but their status as intellectuals doesn't give their views in subjects in which they are not experts any more clout than those of a random person off the street.

The lion's share of this book will deal with myth-busting: analyzing various claims that science and religion have historically been at odds with each other. The purpose of this is to break down the reader's presuppositions about science and religion by having first a little, then a lot, then the very central pillars in how they perceive science and religion, torn down. This is necessary for two reasons: First, the misconceptions here are so numerous and so deeply entrenched that it is impossible for any challenge to them to get a hearing. We may be willing to grant that a few errors have crept into the sacred story of how science beat the snot out of religion, but we think that the general idea *must* be true. Eventually, as

---

4. Wolfe, *Hooking Up*, 133.

5. Sowell, *Intellectuals and Society*, 14; emphasis in original. I'd say something similar about celebrities except they don't have an original field in which they are experts.

6. Herrick, *Introduction to Logic*, 622–23. There's more to it than this, though.

you lose your confidence in this attitude, you'll find yourself left standing in midair above a construction that has collapsed beneath you, à la Wile E. Coyote. At some point you're going to have to look at the camera, blink a couple times, and drop.

Second, there are real issues surrounding science and religion, but they cannot be properly attended to in a climate where simplistic, univocal answers are given that are thought to be the final word. To be sure, plenty of these issues are treated extensively within philosophy—issues such as the nature of causality, mind in a physical world, whether free will is possible, etc.—but when religion is introduced into the conversation, everyone opens their mouths so wide that their brains fall out. You can't even get close enough to ask the deeper questions before the outraged blustering takes over. So before we can delve into the deeper regions we have to clear the air a bit first. This book primarily involves the clearing, not the delving.

After the myth-busting, I'll present a short history of how science and religion have interacted. This is to deal with several issues that don't warrant their own chapters, as well as to present a summary informed by the corrections already addressed. I will also go over two important philosophical issues regarding science and religion: the possibility of miracles and the alleged dependence of science on religion. The latter topic will be the most controversial, which is saying a lot, but I don't see how I can reasonably ignore it. In order to deal with the claim that science and religion are opposed to each other, it is necessary to at least introduce the idea that you can't even *have* science without religion. Unfortunately, such a presentation may seem like a defense, an *apologia*. In my defense, a defense is not my intention. So apologies for the apology.

While I sometimes speak of religion in the aggregate, the real focus here is on the Abrahamic religions, specifically the Judeo-Christian worldview, with a smattering of Islam thrown in for good measure. One reason for this is that the focus of most of the issues surrounding science and religion (not all, though) is directed towards these faith traditions. Another reason is that the idea of a holy book being inspired and immutable is a characteristic of the Abrahamic religions, and when those holy books make specific claims about the world, those indisputable claims become very disputable. So, often, we can't just take the issues with science and Christianity and apply them willy-nilly to religion in general.

I am neither a scientist nor a historian, I'm a philosopher who teaches a course in science and religion and am well-versed in the history of philosophy—which, until recently (a few centuries ago), included what would become science. So I'm proficient with the history of science, the history of philosophy, as well as contemporary philosophy, including philosophy of science and philosophy of religion. Additionally, one of my master's degrees is in Christian theology, and another focused on Islamic philosophy (at least my thesis did). I'm not giving my academic background to establish the claims I make as correct—that's what the bibliography is for—but because I know some readers will want to know. This book constitutes slightly more than half of the course I teach on the subject. Because it's based on my teaching style, it is, perhaps, a little more informal than some academic books. Actually, it may be a little more than a little more.

Nothing I write here is at odds with mainstream historical thought on the subjects in question, and the only points that may seem at odds with mainstream scientific thought are, again, with regards to their philosophical interpretations not the actual scientific claims. Naturally, if the reader finds anything I say here objectionable they are encouraged to research the matter for themselves. As noted, a large part of my goal here is to spark interest in further study of the relevant subjects.

Obviously, the fault for the science vs. religion meme does not lie entirely on one side, not even close. While the primary purpose of the present work is to correct misunderstandings going one way, I also give some examples going the other, of how religious people feed into this cultural divide between religion and science. Often this takes the form of the religious allowing the nonreligious to define religious doctrine. This, it should go without saying (but doesn't—that's why I'm saying it), is bizarre. Numerous Christians do not look to church history or theology or even the Bible when formulating their beliefs, but rather look to what secularists claim the Bible says. They let non-Christians tell them what they have to believe. Since the non-Christians in question are sometimes deliberately interpreting, or even manipulating, the Bible to force a conflict between it and contemporary science, it is, at the very least, *ill advised* for Christians to embrace those interpretations.

A few more points: First, I sometimes call the precursors to modern science "natural philosophy," which is the standard term, or "protoscience," which is my own. More often, though, I just call it "science," even though it wasn't what we would consider science. This is more for ease of understanding than anything else, since it refers to what eventually developed into science. Second, since this book is based on my class

I tend to reference the textbooks I use in the course overmuch. These include Philip Sampson, *Six Modern Myths About Christianity and Western Civilization*; Ronald Numbers, *Galileo Goes to Jail and Other Myths About Science and Religion*; and James Hannam, *God's Philosophers: How the Medieval World Laid the Foundations of Modern Science*. I also overuse two specific online sources, *The Stanford Encyclopedia of Philosophy* and *The Catholic Encyclopedia*. So my reference list should not be taken as anything near exhaustive. Plenty of the biggest names in the field go unmentioned because I don't need to reference them in the course of the course. This includes such lights as Ian Barbour and John Polkinghorne, who are only mentioned in the present sentence, and just so that I can include their names in the index. Third, I have a modest conflict of interest in that James Hannam and I (and Humphrey Clarke) wrote a blog together called Quodlibeta (http://bedejournal.blogspot.com). Humphrey and I aren't posting on it anymore, but James still does. I reference a few blog posts in the footnotes. I also reference another book by James on the history of flat earthery in chapter 2.[7]

Finally, some parts of this book are based on essays I've published. These include "The Myth of Mortification: The Cosmic Insignificance of Humanity and the Rhetoric of 'Copernican Revolutions,'" *Theology and Science* 11 (2013) 289–303; "Cosmic Proportions and Human Significance," *Scientia et Fides* 10 (2022) 263–78; and "Extraterrestrial Soteriology," *Scientia et Fides* 11 (2023) 63–78.

***

Now for the traditional thanking of the thankworthy. Foremost, and this may be obvious, I thank God for sustaining me throughout the whole thing (where "the whole thing" can be taken as either the writing of this book or my entire life). Several people proofread this book, in part or whole, and I offer them my thanks as well and have done my best to heed their suggestions. These include Steven Duncan, James Hannam, Paul Herrick, Jacob Longshore, Matt Savage, and Patrick Tomassi. I also thank the University of Portland for allowing me to corrupt the minds of the youth as well as the corrupted minds themselves for helping me to hone the contents of the class, and hence this book, via their questions. I also thank the librarians at the Clark Library for tracking down so many references for me. Librarians are superheroes.

7. Hannam, *Globe*.

I dedicate this book to four people. First, my wife Krista. She is the rarest of all creatures, someone who is intrinsically cheerful, outgoing, and optimistic, yet *isn't annoying*. I didn't know this was possible, she's almost a logical contradiction. I'm intrinsically cynical, antisocial, and depressed, so being allied with someone of the opposite bent has made everything else possible. Thus I dedicate this book to her, as she is a necessary, but not sufficient, condition for its existence.

Second and third, I dedicate this book to my wonderful, wonderful son Joah and my wonderful, wonderful daughter Kira. Both of them evince a love for science and I want to give them the resources to enjoy it without having to worry about people, from both sides, who say you're not allowed to love science if you're a Bible-thumping Jesus freak. Joah loves biology and chemistry, while Kira is all in for geology. So far.

Fourth is my sister Riki, who was my companion throughout much of my early life (at least the parts I haven't blocked out). She and I were almost like twins. Years ago I had the idea of writing this book and dedicating it to her to guilt-trip her into reading it. Unfortunately, she is no longer with us. So I dedicate this book to her *in memoriam*. I love you forever, Riki.

Chapter 1

The Boring Introductory Chapter

"Myth" has many definitions, some technical, some not. On the technical side, it has to do with traditional stories about the sacred, as distinct from folklore and legend, which don't have (or at least don't require) such an element. It can also refer to a worldview or metanarrative, a culture's *über* story that explains all the other stories. In this sense, "myth" does not convey falsehood, at least not necessarily: a worldview could be true. There are more definitions, and many of them overlap. The primary use of "myth" in the present work is none of these, however; it is, rather, the more colloquial "urban legend." These are just stories or ideas that are passed around in society and are believed by some people, but that are not true: an urban legend is false by definition. If it were true, it wouldn't be an urban legend, it would be a fact.

I. URBAN LEGEND 1.0

For our purposes, I'll divide urban legends into two types. The first is a false story or idea that has no relevance beyond itself. So, for example, many of us were told as kids that it was dangerous to go swimming until an hour or so after eating. This is a filthy lie: you can eat *while* swimming if you want and don't mind getting your beef Wellington soggy. Another one is that the nursery rhyme "Ring Around the Rosy" has survived since the Middle Ages and refers to the Black Death. The title supposedly refers to the particular kind of lesion the plague caused in its victims; posies were to mask the smell of the disease or were brought to place on the grave; ashes refer to how the dead, and sometimes their homes, were burned to try to prevent the spread of the plague; and "we all fall down" because the plague's going to kill all of us eventually. And little kids sing it because, I don't know, children are especially interested in epidemiology or something. In reality, it's just a meaningless word salad that goes back to the mid-nineteenth century and was probably based on an earlier German nursery rhyme that uses similar sounding words (*Ringel, Ringel, Reihe*) but has a completely different meaning.

I can't blame anyone though, since I've believed my fair share of urban legends, and no doubt there are plenty I still believe, many of which I'll never correct. For example, probably around the age of twenty I thought about porcupines for the first time since I was a kid, and said to myself, "Porcupines, right, their hairs are hard quills that they shoot at predators . . ." and I trailed off. *They shoot their quills?* What, do they have a little bit of propellant in every hair follicle? Seriously? Can they shoot individual quills or only clusters of them? Do they aim? I mean, nature can be really, *really* weird, but that didn't sound like the right kind of weird. So you know what happened next: I went to the library (this was before the age of the Internet), looked up porcupines, and found that the idea that they shoot their quills is an urban legend. And I am confident many of you reading this are discovering this fact for the first time.

Another one that is even more embarrassing is cow tipping. I'd heard many times that cows sleep standing up, and people—for amusement, I guess—would sneak into fields at night and run into a cow at full speed trying to knock it over. I believed this. I also believed that cows *don't* sleep standing up but lying down, as I'd seen countless cows doing just that, usually while I was yelling "Moo!" out a car window. Obviously these two beliefs contradict, but for whatever reason I never brought them together, even though they're right next to each other conceptually. I had to read a list of common urban legends to discover that cow tipping isn't a thing.[1] I had the resources to recognize this my whole life but didn't. And again, we all have conflicting beliefs like this that are blatantly, screamingly obvious but we are just too blatantly, screamingly oblivious to recognize it.

---

1. A friend of mine said that cow tipping sounds more like a rural legend than an urban legend. I asked him if I could include that in this book, and he said only if he got royalties. So, Pat, I owe you a dollar.

There are plenty of urban legends that are almost universally believed, like that we have only five senses (sight, hearing, taste, touch, smell). This comes from Aristotle in the fourth century BC. A *sixth* sense is then taken to refer to some paranormal phenomenon like ESP or "I see dead people" or whatever. But there are plenty of senses beyond Aristotle's five. When people refer to their sense of balance, they're not using the term "sense" metaphorically; it's an actual physical sense, no less than sight and hearing, called equilibrioception. Another one is proprioception, sensing where our body parts are when we're not looking at them.[2] This is why when police pull you over for drunk driving—and I strongly discourage you from testing this—they make you stand on one foot, or walk in a straight line, or close your eyes and touch your fingertips to your nose. They're assessing whether you've consumed enough alcohol to affect your equilibrioception and proprioception. (Just to be clear, this is what I've *heard* they do.)

Believing urban legends doesn't mean you're inherently stupid or gullible. This is so even if a particular urban legend is stupid and you were gullible in believing that one. Everyone believes plenty of them, the smartest people in the world believe stupid ones, it's simply part of the human condition. In my own case, I've found that I'm especially susceptible to urban legends that masquerade as corrections of other alleged urban legends (I fear this is because it's a cheap way to feel intellectually superior to others). And, of course, we have an obligation to correct as many of them as possible. The fact that we can't be perfectly rational doesn't mean we shouldn't try to be as rational as we can—anymore than the fact that we can't be perfectly moral means we shouldn't try to be as moral as we can.

The point, then, is this: do your best to believe truths and not believe falsehoods, but be *gracious* towards yourself and others. At some point in your life, probably at many points, you're going to say something stupid and reveal to a roomful of people that you believe porcupines shoot their quills or something and you will be mocked and laughed to scorn. I'm not saying you have to laugh along with them and become friends, just that when it does happen, do your best to shrug your shoulders and move on.

---

2. Oliver Sacks relates a case of a woman who lost her proprioception, equating it to going blind (*Man Who Mistook His Wife for a Hat*, 43–54). And Michael Flynn has a great short story about extraterrestrials who lack proprioception ("Common Goal of Nature").

II. URBAN LEGEND 2.0

I said there were two types of urban legends, the first one being those that have no relevance beyond themselves. They are merely isolated, widely believed falsehoods. The second type of urban legend is a false story that refers to a larger claim that might be true. But just because the lower-level claim is false, it doesn't follow that the higher-level claim it's pointing to is also false. It might be, but it might not be. The problem is that we're often emotionally invested in the larger claims, and the stubbornness and defensiveness we feel when being corrected about things that don't matter kicks into overdrive when being corrected about things that do.

Here's one example: "rule of thumb" supposedly refers to how men used to be allowed to beat their wives with a stick as long as the stick was no wider than their thumbs. This is an urban legend.[3] While we don't know what it originally meant,[4] the beat-your-wife interpretation came around only long after the phrase had been in common circulation. But what's the larger claim that it is being used to illustrate? That some men have mistreated women, and this mistreatment has often taken the form of physical violence. That's true. So even though the urban legend being proposed is false (since urban legends are false by definition), the larger truth it is being used to illustrate is true.

Here's another one that does double duty: European settlers allegedly tried to wipe out Native Americans by giving them blankets infected with smallpox. But historians have been able to find only one case of a single British officer during the siege of Fort Pitt in 1763 trying to kill two specific Native American delegates this way—and it was unsuccessful. At the same time, a British general asked a colonel whether it was possible to spread smallpox among the Native Americans to end the siege, although blankets were not mentioned and they never actually tried anything as far as we can tell.[5] Moreover, the reason the attempt in the first case failed is because it's all but impossible to catch smallpox in any way other than via direct contact with another person who has it. Blankets won't work. It's theoretically possible but we have no evidence that anyone has ever caught smallpox except by direct exposure to other people with smallpox.

---

3. O'Conner and Kellerman, *Origins of the Specious*, 123–26.

4. When I was a kid I thought it referred to a painter looking at his artwork with his arm outstretched and his thumb up before his eyes like a kind of rifle sight and thinking, "Eh, good enough."

5. Ranlet, "British, Indians, and Smallpox."

Thinking you can catch it from a blanket is like thinking you can catch an STD from a toilet seat.

So this story is false, but what are the larger claims it's pointing to? First, some Europeans and their descendants in the Americas mistreated Native Americans. That's true. We can certainly debate the extent of it and how representative it was, but it happened. Second, that Europeans brought diseases to the Americas that wiped out many of the Native Americans. That's also true, although obviously it was not intentional.

Anyway, you can see why people could be offended more easily when told they believe an urban legend when it is one of the bricks they use to establish their worldview. They may not immediately see (or be willing to see) that challenging the truth of a particular story does not necessarily challenge the truth of the larger story. It's easy, though, to take a challenge to the smaller claim as automatically being a challenge to the larger claim. But as long as the larger claim's truth can be established independently of the urban legend(s) used to illustrate it, there's no problem.

Sometimes, however, the larger claim's credibility *is* tied up with the truth of the smaller claims given in support of it—namely, when the larger claim is just a generalization of all the smaller claims. In this case, if you refute enough of the smaller claims, you essentially refute the larger claim as well. That's what a large part of this book is: looking at urban legends that illustrate the larger claim that science and religion are at war (not to mention that science is winning the war or has won it). The truth of this claim is dependent on the examples that allegedly illustrate it, and if those examples are false, the larger claim, which is *ex hypothesi* a generalization from the examples, is false as well.

Occasionally, when an urban legend—or a misunderstanding or lie—that illustrates a larger claim is discovered to be false, some people respond by saying, "OK, well, that particular claim may not have been true, but it made people more aware of the larger issue it was illustrating and that's more important." For the love of God, *don't ever do this*. If the truth doesn't allow you to make your point as forcefully as you'd like, too damn bad. Give correct illustrations of your larger point, and if you find out an illustration is not correct, apologize and find another one that *is* correct. (Also: don't qualify the apology by emphasizing how you still disagree with your opponents. We already know this. Just apologize, pour some ashes over your head, and move on.) Merely on a practical level, if people realize that truth takes a back seat to your promotion of a larger claim, they will not trust your judgment in general and about the larger claim in particular. In fact, some who agree with you about the larger claim might start questioning it.

More importantly, though, it's immoral. When you condone using falsehoods to promote a larger claim, you are manipulating people and disrespecting their autonomy by using them as tools to your own end. In doing this

> we are making use of people, trying to bypass their understanding and judgment to trigger their will and possess them for our purposes. Whatever consent they give to us will be uninformed because we have short-circuited their understanding of what is going on.... [We should] respect the soul need of human beings to make their judgments and decisions solely from what they have concluded is best. It is a vital, a biological need. We do not thrive, nor does our character develop well, when this need is not respected.[6]

We tend to be less critical of stories that illustrate a larger claim that we believe and think important, and more critical of stories that conflict with it or illustrate a larger claim we disagree with. This is called confirmation bias. By itself, this need not be irrational: stories that support something we believe will "fit" in with what we already (think we) know about the issue, and so cohere with it, and coherence is often an important element in deciding a claim's truth.[7] Stories that do not cohere or that go against it do not fit in with the larger store of knowledge that we (think we) have, and so are not explicable by the overarching theories by which we explain things.

Having said that, obviously we need to do our best to apply the same standards to both sides, whether they let us believe what we want or not. We have the biases we have, but we must leave enough room for information to *change* our biases by being true yet inconsistent with them. And while we're inclined to apply this to the people we disagree with, the point is that we need to apply it to ourselves.

III. HORSE SENSE

To ease you into the myth-busting portion of our program, I'm going to start with some urban legends about science and religion most people haven't heard of, so rejecting them won't be too distressing for those who

---

6. Willard, *Divine Conspiracy*, 174–75.
7. There's an entire theory of knowledge called coherentism based on this.

believe there is a conflict between them. In general, a lot of the urban legends about science and religion contrast the ancient and medieval worlds with the modern world and the Scientific Revolution. The Middle Ages were often called the Dark Ages to contrast them with the Renaissance (which means rebirth) and the Enlightenment (which means ungloomification). Sometimes, "Dark Ages" was just used to refer to the early medieval period, and particularly the dearth of information we had about it—it's "dark" because we can't see it—but more often it was used as a value judgment about how *they* were in the dark in terms of knowledge, and was eventually extended to cover the entire medieval period. Historians don't use the term "Dark Ages" much anymore.

Sometimes the contrast between the premodern and modern worlds is expressed by showing the absurd intellectual methods and interests the medievals had, and which were in place specifically because their religious beliefs led them astray.[8] For example, take this quote, allegedly from Francis Bacon, one of the giants of the Scientific Revolution:

> In the year of our Lord 1432, there arose a grievous quarrel among the brethren over the number of teeth in the mouth of a horse. For thirteen days the disputation raged without ceasing. All the ancient books and chronicles were fetched out, and wonderful and ponderous erudition such as was never before heard of in this region was made manifest. At the beginning of the fourteenth day, a youthful friar of goodly bearing asked his learned superiors for permission to add a word, and straightway, to the wonderment of the disputants, whose deep wisdom he sore vexed, he beseeched them to unbend in a manner coarse and unheard-of and to look in the open mouth of a horse and find answer to their questionings. At this, their dignity being grievously hurt, they waxed exceeding wroth; and, joining in a mighty uproar, they flew upon him and smote him, hip and thigh, and cast him out forthwith. For, said they, surely Satan hath tempted this bold neophyte to declare unholy and unheard-of ways of finding truth, contrary to all the teachings of the fathers. After many days more of grievous strife, the dove of peace sat on the assembly, and they as one man declaring the problem to be an everlasting mystery because of a grievous dearth of historical and theological evidence thereof, so ordered the same writ down.[9]

---

8. Like asking how many angels can dance on the head of a pin, a phrase that was invented in the modern era to mock the medieval philosophers.

9. Quoted in Munn, *Introduction to Psychology*, 4.

You can guess where I'm going with this. This text has been traced back to 1901 (which is several centuries after Francis Bacon, in case you were wondering) in an article in the *Machinists Monthly Journal* that claims to be repeating a passage "from the Chronicle of an Ancient Monastery."[10] The story it relays never happened. More than that it never *would* have happened: the medieval theologians would never have engaged in such a farce. The kernel of truth—and it's just a kernel—is that a lot of premodern thought was strongly rationalist as opposed to empiricist. Roughly, rationalism says that the best way to discover truth is by thinking, while empiricism says that the best way to discover truth is by looking. Obviously, it depends on what truth is in play but it's not always clear cut. The error of the medieval thinkers is that they put some things in the rational basket that should have gone in the empirical basket, but how many teeth a horse has was not one of them. And in my megalomaniacal opinion, plenty of thinkers in our culture put some things in the empirical basket that should go in the rational basket.

IV. COMETS AND CONCERNS

Here's a more detailed example. In June 1456 Halley's Comet appeared in the sky and was seen as a bad omen. So when, in early July, the Turks began the siege of Belgrade, it was seen as a confirmation of the comet being a potentially potent portent. Halley's Comet so disturbed Pope Callixtus III that he resorted to drastic measures: he excommunicated it.

For years this story was told to illustrate how absurd and superstitious religion is, especially when contrasted with science. Carl Sagan expressed this contrast by titling one of his books *The Demon-Haunted World: Science as a Candle in the Dark*, which includes a chapter entitled "The Fine Art of Baloney Detection."[11] He even refers to Callixtus excommunicating Halley's Comet in a book he coauthored with his wife, Ann Druyan, which was published to coincide with the appearance of Halley's Comet in 1986.[12] The story, however, is false, an urban legend. It was popularized in 1796 by Pierre-Simon Laplace, who got it in turn from Bartolomeo Platina's *Vitæ*

---

10. *Monthly Journal of the International Association of Machinists* 13 (1901) 129–30, https://digitalcollections.library.gsu.edu/digital/collection/IAM/id/42362/rec/67, found via https://dsimanek.vialattea.net/horse.htm.

11. Philosophers aren't as delicate; see Frankfurt, *On Bullshit*.

12. Sagan and Druyan, *Comet*, 26–28.

Pontificum, which was written in 1476, just twenty years after the events, which is pretty close in historical terms. Platina wrote,

> A maned and fiery comet appearing for several days, while scientists were predicting a great plague, dearness of food, or some great disaster, Callistus decreed that supplicatory prayers be held for some days to avert the anger of God, so that, if any calamity threatened mankind, it might be entirely diverted against the Turks, the foes of the Christian name. He likewise ordered that the bells be rung at midday as a signal to all the faithful to move God with assiduous petitions and to assist with their prayers those engaged in constant warfare with the Turks.[13]

One immediate takeaway is that Platina does not mention anything like an excommunication. He doesn't even directly connect Halley's Comet to Callixtus's decree, although he obviously implies it by putting them right next to each other. The decree, *Cum His Superioribus Annis*, exists but does not mention Halley's Comet—it was released a few days before the siege of Belgrade began, but was imminent. Platina may have implicitly tied the comet and the siege together, but there's no evidence that the pope, or anyone else, did.

When Laplace wrote about the story several centuries later, he claimed (erroneously) that Callixtus ordered prayers of condemnation be directed towards Halley's Comet, which was translated into English as saying he anathematized it.[14] This was taken as claiming that Callixtus exorcized the comet, and some writers switched from "exorcized" to "excommunicated" because all those religious terms mean the same thing anyway. This despite the fact that *only members of the Catholic Church can be excommunicated from it* since it means to remove a member of a congregation from said congregation. For Halley's Comet to be excommunicated it would have to have been a member of the Catholic Church, which (spoiler) it wasn't. Only people can be members of a church, not inanimate objects. Even if you were to suggest that some people might have thought there were nefarious spiritual agents associated with the comet, those agents would not have been members of the Catholic Church and so not eligible for excommunication. Thus the very idea of a pope excommunicating a comet is absurd, even though some very smart people believed it.

---

13. Translated and quoted in Stein, "Bartolomeo Platina."
14. "Calixte ordonna des prières publique, dans lesquelles on conjurait la comète" (Laplace, *Exposition du système du monde*, 60; *System of the World*, 62).

Is there a kernel of truth here too, though? Weren't people in the fifteenth century more superstitious, believing in astrological signs and the like? Well, I *assume* so, although I've never researched it. Yet we still have astrology. Most newspapers print horoscopes every day. And astrology itself is not a Christian nor even a religious phenomenon: there certainly have been, and are, plenty of Christians who believed it, but astrology is present in nearly every civilization we have evidence for. If medieval people did fear comets as evil omens, it was not because of Christian beliefs or the words of the Bible or church tradition. It would have been because of the cosmology of the day, traceable back to Ptolemy and Aristotle, which held that, above the moon, the universe was perfect and immutable. Comets appear to be further away than the moon and do not seem to be part of a pattern, so they could be seen as imperfect alterations in the perfect, inalterable universe. But these concepts did not originate in religion, nor were they specifically religious concerns, much less Christian concerns. The closest thing to signs in the sky in the Bible are *good* omens (like the star of Bethlehem), not bad.

Despite this, Sagan and Druyan portray the concept of comets among the ancient Greeks and Romans as generally sober reflections and medieval Christians as imposing superstitious forecasts upon them, "with hardly ever, even as a passing thought, the view that comets might be just a part of nature."[15] In actuality, it was Christians who led the way in understanding comets as natural phenomena,[16] and the official teaching of Christianity was to oppose astrology, at least insofar as it implied that we are not free and so cannot be held responsible for our actions.[17] Regardless, after disparaging medieval Christianity's views on comets, Sagan and Druyan detail "the story most astronomical writers relate about Callixtus and the comet" as their only example demonstrating it.[18] In the end, however, they conclude that it's not true (so they got that going for them, which is nice), but that leaves them without any justification of their claim. It's a little weird. I mean, if you're making a point, and you provide a single piece of evidence supporting it, and then admit that the evidence is invalid, you haven't *made* your point; you've only *asserted* it.

The basic idea behind astrology is that, since big changes in the sky correspond to big changes on earth (which stars rise above the horizon

---

15. Sagan and Druyan, *Comet*, 26.
16. Sampson, *Six Modern Myths*, 17.
17. Hannam, *God's Philosophers*, 126.
18. Sagan and Druyan, *Comet*, 27.

indicates the change of seasons, when you should plant your crops), small changes in the sky should correspond to small changes on earth. This was a reasonable inference, it was simply an incorrect one. As noted, Christianity condemned astrology when it was used to suggest that we are all locked into destinies over which we have no control instead of being moral agents with free will, but it was usually allowed (but still often rejected) when it was considered only as a potential guide for predicting future physical events. And although astrology is often thought to be a bastardization of astronomy, astronomy evolved out of astrology. This was common: the genuine sciences often started as something we consider pseudo-scientific but then eventually developed into authentic sciences. In the same way, alchemy, the attempt to turn iron or lead or some other base metal into gold, was not a vulgar corruption of chemistry. Alchemy came first and evolved into chemistry. It was the medieval alchemists who first isolated the three mineral acids as well as alcohol.[19]

V. NULL HYPOTHESIS, *OR* NADA PROBLEM, *OR* ZERO TOLERANCE, *OR* TALES OF A FOURTEENTH-CENTURY NOTHING

Here's another story from the science vs. religion cage match that you probably haven't heard. When medieval Europe came into contact with Islamic learning, they discovered that the Muslims used a symbol for the number zero. There is no symbol for zero in Roman numerals. So there was immediate resistance to it: not only was it different, it was foreign and *unchristian*. But more than that, a symbol is a thing, it's *some*-thing. Zero, however, is *no*-thing. Using something to represent nothing ascribes being to nonbeing, it elevates nothingness into a category of existence. This is not merely misleading or inappropriate, it's flat out evil. In the tenth century Gerbert d'Aurillac, before becoming Pope Sylvester II, used a symbol for zero and was accused of trafficking with evil spirits, and was forced to repudiate it. The church, at least on a local level, tried to ban the number zero numerous times.[20]

---

19. Hannam, *God's Philosophers*, 132.
20. This story was popularized in Dantzig, *Number*, and it has been repeated in many books since, such as Seife, *Zero*.

Given the current theme, it won't surprise you to learn that this is an urban legend. There is no evidence that any of it is true. The church never banned or did anything similar regarding the number zero, either locally or churchwide. Nor were Muslims the first to represent zero: many other cultures prior to the advent of Islam had some symbol (sometimes just a space) to stand for zero.[21] Roman numerals do not have a symbol for zero, but within the Roman Empire symbols were used to represent zero in Ptolemy's day (second century AD), and no one batted an eye.[22]

One probable source of this urban legend is that Greek philosophy debated whether the concept of "nothing" even made sense and so whether we can refer to it. This skepticism of absolutely nothing comes from Parmenides[23] and was applied to mathematics (zero) as well as physics (vacuum). Many people in the ancient and medieval West believed that a vacuum could not exist. And just as a vacuum is the absence of matter, so zero is the absence of quantity. However, it was taken as a scientific and philosophical claim, not a theological or religious one. The only contribution theology made was to say that although a vacuum could not exist *naturally*, if God wanted to create one *supernaturally*, of course he could.[24]

Another source to the Zero Myth is that there was some economic and political resistance to Arabic numerals (not theological or religious resistance) when they were introduced into Europe. This was not because of their inclusion of a symbol for zero, nor was it because they were foreign or of Islamic origin, it was because they created problems in the economic sphere and led to widespread fraud. The Florence city council—not the church—banned the use of Arabic numerals around 1300 because of this and the economic havoc it was causing.[25] But this was a ban on all Arabic numerals, not just zero, and had nothing to do with religion, theology, or the church.

A further reason for resistance to Arabic numerals came from professional abacus users (abacusers?). Abacuses, or abaci, are those frames with wires stretched across them and beads on the wires that are shifted around so that they function as manual calculators. Roman numerals do

---

21. Barrow, *Book of Nothing*, 15–48.
22. Neugebauer, *Exact Sciences in Antiquity*, 13–14 and plate 2.
23. Copleston, *History of Philosophy*, 1:49–50; Ring, *Beginning with the Pre-Socratics*, 101–3.
24. See below, 187.
25. Barrow, *Book of Nothing*, 45.

not lend themselves well to mathematical calculations, so "for basic arithmetic, it was far easier to use an abacus but this involved a completely different system of counting."[26] But the new Arabic numerals made arithmetic much easier and accessible to the masses, taking away the source of income from those who used abacuses to make their living. So the abacusers had no more abacustomers. This led to a lot of abacussing.

VI. THE LARGER MESSAGE

Both the Comet Myth and the Zero Myth are pointing to the larger claim that science and religion are at odds with each other. It's more complicated than that though: they're making an intermediate point that itself is one of the claims that support that larger claim. This doesn't mean that people who believe or spread the Comet Myth and Zero Myth are themselves trying to make any intermediate or larger claim. Charles C. Mann repeated the Zero Myth in a book,[27] and some bloggers challenged him about it. After some polite back-and-forth, Mann recognized the error and corrected it in the revised edition. He didn't have any axe to grind because he wasn't trying to defend any larger claim.

Anyway, what's the point, the intermediate point, that the Comet Myth and Zero Myth are trying to make? Well, the Zero Myth is meant to show how religion leads people to reject obvious truths. It shows how *closed* religious people are to accepting ideas, to the point of absurdity— new ideas, *scientific* ideas, in particular. A mathematical advancement is made, but because the Christians didn't make it, they were resistant and even hostile to it. Their religious beliefs made them think the new idea is not just wrong but spiritually dangerous.

Here's a similar example: William J. Murray is the son of Madalyn Murray O'Hair, who was an outspoken and (perhaps) the most famous atheist in the second half of the twentieth century. According to Murray, his mother once spoke at Harvard University and told the assembled students and faculty that nobody realized that sex leads to pregnancy until the nineteenth century. That's the nineteenth century AD.[28] The reason for this ignorance is that religious people are terrified of sex and science. Hopefully, the reader will know that O'Hair's claim is false, like really,

---

26. Hannam, *God's Philosophers*, 158–59.
27. Mann, *1491*, 19, 215.
28. Murray, *My Life Without God*, 222.

extremely, holy-crappily false, even though it was accepted without question by her throng of "skeptical" followers. But her claim was trying to make the same point that the Zero Myth is making: religion leads people to reject true and even obviously true scientific ideas.

What about the Comet Myth? That's trying to make the converse point. Instead of saying how closed religious people are to accepting true ideas to the point of absurdity, the Comet Myth is saying how *open* religious people are to accepting *false* ideas to the point of absurdity. That's why "one astronomical writer after another" accepted it.[29] The irony is that these scientists uncritically accepted a goofy urban legend because they thought it showed how *other* people were gullible.

I have a parallel story here too. In 1849 Samuel George Morton measured the volume of skulls of various races, discovering that white skulls (that is, skulls of Caucasians—all skulls are white unless they still have blood on them) had larger average capacities than black skulls.[30] In 1981 Stephen Jay Gould published a book, based on an earlier essay, where he argued that Morton had falsified his measurements.[31] Specifically, he filled the skulls with ball bearings, then counted how many of them had fit into each type of skull. But Gould wrote that Morton had packed down the ball bearings in the white skulls to ensure that his measurements would yield the largest volume. Gould's point was that even the most elementary parts of science were not free from bias, even something as basic as volume measurement. Morton's racism found its way into his research, and while this doesn't have any intrinsic connection with religion, it's a broader point about false and outmoded ideas twisting science. Just as Morton's biases contaminated his studies, so religious beliefs can contaminate other studies, and these contaminations can be present at the most basic level of tabulation of averages and measurements.

In the twenty-first century, however, some enterprising young scientists did something radical: *they remeasured the specific skulls that Morton had used*. It turns out that Morton's measurements were pretty accurate, and the few inaccuracies imputed greater skull capacity to the African skulls, the opposite of Gould's claim.[32] Moreover, Gould assumed that Morton equated intelligence with skull capacity, and tried to ensure that the white skulls were bigger to prove the superiority of white people.

29. Sagan and Druyan, *Comet*, 27.
30. Morton, *Crania Americana*.
31. Gould, *Mismeasure of Man*.
32. Jason E. Lewis et al., "Mismeasure of Science."

Morton certainly did think the white race "is distinguished for the facility with which it attains the highest intellectual endowments,"[33] but in measuring skull capacities, he was actually investigating a different issue: whether the various human races had separate origins (polygenism) or were all descendants from a single set of ancestors (monogenism). The latter had always been taken as the biblical position in the Judeo-Christian tradition, but it was being challenged in the first half of the nineteenth century. The large part of the motivation for the polygenist movement was racism, but two of its most important advocates, George R. Gliddon and Josiah C. Nott, were also strongly motivated by their contempt for religion and Christianity in particular with its affirmation of the unity of all humankind.[34] Indeed, throughout the late nineteenth century and into the early twentieth, polygenism was often considered a greater threat to Christianity than evolution.[35]

Again, Morton did ascribe greater intelligence to white people, but his point here was to challenge monogenism. This may have affected which skulls he chose as representatives of each ethnicity,[36] which would have been sufficient to establish Gould's point that Morton's racist beliefs found their way into his scientific endeavors. But it didn't affect the aspect of his endeavors that Gould specifies, namely his measurements and tabulations of averages.

Gould's larger point, or one of them, is that scientists are human, and just as prone to biases as us lesser mortals, and of course this is true. The fact that Gould accidentally made this point by being the example of it himself (rather than the person he was pointing to) doesn't detract from it; it only makes it funnier. In the same way that "most astronomical writers" gullibly repeated an urban legend to show how gullible other people were, so Gould let his bias color his research about how bias colors other scientists' research. Thus, we do alchemy in reverse, turning Gould into irony. I would respond that the suggestion that bias in science is as widespread as he implies is exaggerated. Science is fallible, of course, but part of its glory is its self-correcting nature.

33. Morton, *Crania Americana*, 5.

34. Blair Nelson, "Men Before Adam," in Lindberg and Numbers, *When Science and Christianity Meet*, 161–82 (esp. 169–70).

35. Nelson, "Men Before Adam," in Lindberg and Numbers, *When Science and Christianity Meet*, 173–80.

36. Nelson, "Men Before Adam," in Lindberg and Numbers, *When Science and Christianity Meet*, 167.

The reason why people could be taken in by such a silly story as a pope excommunicating a comet is because it fits with their views on the nature of science, the nature of religion, and the relationship between the two. Madalyn Murray O'Hair and her followers were completely contemptuous of religion and religious believers, so any claim that justified this attitude—even a patently absurd claim like that no one ever realized that sex led to pregnancy until the nineteenth century—was plausible to them. They needed to be more skeptical of their skepticism. So should we: we need to be aware of what we want to believe and what we don't want to believe so that we can see how these motives influence us. This doesn't mean we should try to force ourselves to *stop* believing what we believe, but to examine how what we believe affects our willingness to accept or reject other beliefs.

Chapter 2

The Shape of Things to Come

I. SPHERE FACTOR

PROBABLY THE MOST WELL-KNOWN example of absurd religious beliefs getting in the way of science is how people thought the earth was flat before Columbus. Christian Europe believed this because a flat earth is part and parcel of the biblical worldview, and because church tradition demanded it. To challenge the claim that the earth is flat was to challenge the authority of the church and the truth of Christianity. When Columbus suggested that the earth might be round[1] he was immediately denounced and even threatened with charges of heresy by the bishops and inquisitors, since they didn't want anyone defying their authority to dictate what was real and true. Fortunately, he convinced some monarchs that he might be right, so they supplied him with a few small ships and a crew, and the rest is history.

Except, apart from the funded by monarchs part, *none* of that is history, it's urban legend. By the time Columbus sailed the ocean blue, everyone in Western civilization already knew the earth was round. In fact, they had known it for nearly *two thousand years* by Columbus's time. The first direct mention of a round earth is in *Phaedo*, one of Plato's Socratic dialogues from the beginning of his middle period.[2] This would either place it about 400 BC when Socrates would have said it or in the

---

1. "Round," here, means "spherical," not a flat circle.
2. Plato, *Phaedo* 97d, 108e.

370s BC when Plato wrote it.[3] But since *Phaedo* portrays it as an issue already being debated in the Greek Empire at the time, the idea predates them. Aristotle, Plato's student, credited it to the Pythagorean school, which probably meant Philolaus, a contemporary of Socrates from the fifth century BC.[4] Several centuries later, Diogenes Laertius claimed that Pythagoras himself suggested a spherical earth in the sixth century BC, but elsewhere, he credits it to Parmenides in the fifth century.[5]

In the fourth century BC, Aristotle and Eudoxus gave scientific arguments for the sphericity of the earth—for example, when one travels south, one can see stars that cannot be seen from the Greek Empire (like Canopus), and other stars and constellations would rise higher above the horizon. Plus, lunar eclipses had been observed, and the shadow of the earth on the moon clearly showed the earth to be a sphere. In the third century BC, Eratosthanes measured shadows in southern and northern Egypt on the summer solstice and noted that objects cast smaller shadows in the south. He concluded that the sun was more directly over the southern lands than the northern, which indicated that the surface of the earth was curved. From his measurements, he calculated the earth's circumference, and, incredibly, he was pretty close. The Stoics took over and spread the gospel of the globe far and wide, among both the literate and the great unwashed. While there was some hemming and hawing, the shape of the earth soon became the nearly unanimous view in Western civilization thereafter.[6]

Then we get to Ptolemy's cosmology in the second century AD which provided a mathematically rigorous model of a view based on Aristotle's cosmology: that the universe was arranged in concentric spheres.[7] Earth is the sphere at the center, and the next sphere up is that of the moon. The sphere of the moon is not the moon itself, it is the sphere in which the moon is *embedded*, and in which it orbits the earth. Above and around that are the spheres of Mercury, Venus, the Sun, Mars, Jupiter, Saturn, and finally the sphere of the stars. The final sphere is the boundary of the entire universe. So the planets and stars, the sun and the moon, moved in circular orbits (the most perfect form of motion) around the center (earth). This was all before telescopes, so they were restricted by

---

3. See below, 88–89.
4. Aristotle, *On the Heavens*, 293a; Hannam, *Globe*, 66–79.
5. Diogenes Laertius, *Lives of Eminent Philosophers* 8.1.48; 9.3.21.
6. Hannam, *Globe*, 80–112.
7. Ptolemy, *Almagest*.

what could be seen by the naked eye, thus Uranus and Neptune were not included.[8]

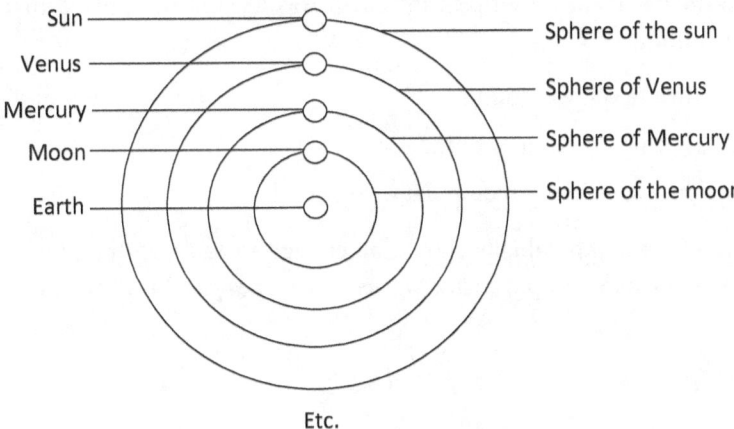

This cosmology was all but universally accepted in antiquity and the Middle Ages, and it required the earth to be spherical. The sphericity of the earth was asserted in Boethius's *Consolations of Philosophy*, the Venerable Bede's *Natural History*, Sacrobosco's *De Sphaera Mundi*, the *South England Legendary*, Metz's *L'image du monde*, Mandeville's *Travels*, Dante's *Divine Comedy*, and elsewhere. A spherical earth was assumed just about everywhere the issue came up throughout the Middle Ages, based on Ptolemy's cosmology.

In addition, they *portrayed* the earth as a sphere. The globus cruciger is an orb with a cross on top of it that often formed a part of royal regalia and was sometimes used by medieval kings to kill violent rabbits. The orb represents the earth and the cross on top represents Christ's dominion over it. Some ruler or leader holds the globe to indicate "he's got the whole world in his hands." Globus crucigers and portrayals of them on coins and artistic depictions are present throughout the Middle Ages and even in the modern era, the earliest dating back to the late fourth century AD. Before this, the Roman Empire wasn't officially Christian, so the image didn't include the cross, but was just a globe. Portrayals of someone holding a globe like this go back centuries before the cross was added, and the point was to indicate the scope of a particular leader's

---

8. Although, interestingly, Uranus can be seen by the naked eye (at least by some), but it wasn't recognized as a planet until the late eighteenth century.

power by holding the world (or sometimes having the world under one's foot). Roman coins depicted a spherical earth as a matter of course.[9]

So we have three *very* strong reasons for thinking that the ancients and medievals did *not* think the earth was flat but recognized that it was round.

1. They said it was round.
2. They portrayed it as round.
3. Their cosmology required it to be round.

Each of these individually should be enough to establish the point. So why do so many people, including very smart people, think otherwise?

II. THE ARGUMENTS AGAINST

Maps

One argument that I haven't seen made by historians but have seen deployed by Internet users is that *mappa mundi* (medieval European world maps, such as T-O maps) clearly depicted the earth as a flat circle. So, contrary to point 2 above, the medievals did not portray the earth as round but as flat, at least some of the time.

I've looked for more depth to this argument and haven't found it, although that's not to say there isn't any somewhere. At least on the level I'm presenting it, the response is obvious: *of course* maps are flat. They're maps. We make maps too and it doesn't mean we secretly think the earth is flat. Their prolificity in the Middle Ages and today is simply because maps are easier to make than globes. Moreover, they were partial; they were maps of the *known* world, with the full understanding that there was more world beyond them. Plus, they were made more for artistic reasons than practical ones. You might as well say that since paintings are on flat canvases, they indicate that painters don't believe in depth.

So portraying something in two dimensions doesn't indicate that the thing being portrayed is two dimensional, just that the medium being used to portray it is two dimensional. But if maps aren't portrayals of a flat earth, why are globus crucigers portrayals of a round earth? Because a globe adds something a map doesn't have: a third dimension. If you thought the world was a sphere, there would be very good reasons why

9. Carile, "Globus Cruciger."

you would draw flat maps of it: maps are easier to make, easier to transport, easier to store, *cheaper* to make, transport, and store. If you were making a map of a relatively small area, trying to indicate the curvature of the earth would be unhelpful at best. If you thought the world was flat, however, there would be no reason why you would represent it as a globe.

Antipodes

Another potential objection is the antipodes controversy. Antipodes refers to lands and people who live on the opposite side of the world, specifically who *walk* on the other side of the world. *Pódes* means foot, so antipodes are where people's feet are opposite one's own. The controversy was whether such people (antipodeans or antichthones) existed. Many people denied it because they thought there was no way to get from one side of the earth to the other: either the oceans were too wide to cross or the equatorial zone was too hot to pass through.[10] So if there *were* people on the other side of the world, they couldn't have gotten there from this side and would not share the same origin as us; they would be a separate creation.

The most important point to make here is that the skepticism about antipodes and antipodeans was *anthropological* in nature, not *geographical*. It was about the extent of humanity, not the shape of the earth.[11] For all they knew, there wasn't even any land on the other side of the world and thus nowhere for any self-respecting antipodean to stand on his own two *pódia*. So the whole idea was completely hypothetical.[12]

So why was it controversial? Here there *was* a religious connection: antipodeans raised some theological issues and brought two Christian concepts into tension. The Christian story of humanity is that human beings are created in God's image, but our ancient ancestors sinned against God (the fall) and as a result we, their descendants, are born into a state called original sin, meaning we have an innate tendency to sin. One concept derived from this is that original sin is transmitted biologically, so that it affects only human beings.[13] Pandas, puffins, and pygmy

---

10. Russell, *Inventing the Flat Earth*, 20.

11. Besides which, a flat earth would have an "other side" to it as well. Unless the world was a Möbius strip. Actually that would be pretty cool: a Möbius earth. That reminds me of Arthur C. Clarke's short story "Wall of Darkness." Go read that.

12. The best account of the ancient and medieval speculations about antipodes and antipodeans that I've found is Goldie, *Idea of the Antipodes*, 15–70.

13. Augustine, *City of God* 16.9.

marmosets are not moral agents and so cannot commit sins. However, another concept is that the fall also fundamentally changed the entire earth, possibly the entire universe, so that it doesn't just affect human beings—we are fallen creatures living in a fallen world. Neither of these concepts are explicitly made in the Bible but they are good faith attempts to understand it.[14]

So if there were antipodeans then they would not be fallen because they would not be under the specter of original sin (not being descended from the original human beings who sinned against God). But they *would* be fallen because they would be living in a fallen world. This isn't a contradiction—it would mean they're fallen in one sense and not fallen in another sense—but if they are moral agents, it would imply they are created in the image of God, might not have sinned against him, yet suffer the consequences of *our* sin. This forces us (or at least the Christians among us) to ask, if they're created in the image of God, did their remote ancestors sin against God too, and did it have the same effect? Did the fall of humanity—I mean, the humanity of the northern climes—affect them or do they have their own stain of original sin? Did their fall affect the world too? How? And if they are fallen, does Christ's atonement apply to them? Or was Christ incarnated among them too? Or do they have a completely distinct path to salvation? If Christ's atonement applies to the antipodes, *how* can it do so if they're not biologically related to us? Should we try to evangelize them or would this take them away from the path of salvation God provided for them?

The problem is not that these questions made Christianity less plausible. The problem is that, absent further revelation, there's simply no way to answer them. Attempts to do so were almost entirely speculative. One Christian's sense of how to answer them will differ from another's and they are impossible to nail down with genuine exegetical or theological points. This is common enough: there are plenty of unanswered questions in Christianity like this, such as what is the ultimate fate for those who have never heard of Jesus. If we deny that original sin is transmitted biologically or that the fall instigated a global or cosmic alteration (or both), then it is not so very intractable, but if we grant them, as Christians traditionally have, the Bible did not provide the premoderns with the resources to understand the spiritual status of antipodeans.[15]

---

14. See below, 165–73.
15. I'll address this in more detail in later chapters.

Yet, despite these issues, "there never was a doctrine of the Christian Church condemning the idea that there might be inhabitants of the southern temperate zone or of a presumed fourth continent."[16] It was just easier to deny the existence of antipodeans than to speculate about them. Sometimes this denial was made angrily and forcefully, but the closest anyone ever came to getting in trouble for affirming the existence of antipodeans was when St. Vergilius of Salzburg got on the bad side of St. Boniface. Their first rumble had to do with the exact wording of the sacrament of baptism: a priest had administered it incorrectly, but Vergilius thought it was still efficacious. Boniface disagreed and brought the issue before Pope Zachary, who agreed with Vergilius. After this, Vergilius apparently argued in favor of antipodes. Boniface again brought the issue before Zachary, arguing that the existence of antipodes was contrary to Scripture. The sympathetic pope said if Boniface could prove it, he'd kick Vergilius out of the priesthood and the Catholic Church. Since Vergilius later became the bishop of Salzburg, and was eventually canonized as a saint, we can safely assume he was exonerated.[17]

Anyway, the *point*—I had one—is that the controversy over antipodes and antipodeans wasn't over the shape of the earth but the geographical extent of the human race and the theological questions that arose in light of the possibility of other creatures who *may* be created in God's image, *may* be fallen, and *may* have a distinct form of salvation. None of this would amount to an objection to Christianity, it would just be an unanswerable question. And they thought the earth was a globe throughout.

The Bible

Another objection people raise is that medieval people must have believed the earth is flat because the Bible *says* it's flat. Ask for a specific chapter and verse and you're unlikely to get one, but there are a few passages that can be brought up in this context.

First, a common biblical phrase is "the ends of the earth," for example, Ps 48:10: "Like your name, O God, your praise reaches to the ends of the earth; your right hand is filled with righteousness." A similar phrase is "the four corners of the earth," such as in Isa 11:12: "And He will lift up a standard for the nations and assemble the banished ones of Israel,

---

16. Stevens, "Figure of the Earth," 274.
17. Turner, "St. Vergilius of Salzburg."

and will gather the dispersed of Judah from the four corners of the earth" (NASB).

Just like maps, however, *we* use those phrases too (although they are a bit antiquated in contemporary English), and it doesn't imply that we think the earth is flat. As with words like "sunset" and "sunrise," "this is a perfectly acceptable type of phenomenal terminology, employed by all languages at all periods of their history."[18] Phenomenal language describes things as they appear to us without implying anything more about them. "The four corners of the earth" simply refers to the furthest one can go from one's home or present position in the four cardinal directions: north, south, east, and west. "The ends of the earth" means the same without specifically bringing up the four directions. Isaiah 43:5–7 shows that they're talking about the same idea: "I will bring your children from the east and gather you from the west. I will say to the north, 'Give them up!' and to the south, 'Do not hold them back.' Bring my sons from afar and my daughters from the ends of the earth."

A more explicit passage is in Rev 7, which describes a vision of "four angels standing at the four corners of the earth." This seems to be an actual description of a flat earth not mere phenomenal language. Unfortunately, it's a *vision*. Visions are essentially waking dreams that function as communiqués from God (or sometimes less respectable members of the spiritual world). The vision in Rev 7 is not trying to describe the physical world or its shape.

One could still object that many of the biblical authors would have believed whatever the science of their time and place was, and if that included a flat earth, it would have found its way into the traditional phrases and imagery they used. I see no reason to deny this; or care about it. Contemporary physics teaches us that solid objects consist mostly of empty space, but prior to Ernest Rutherford's model of the atom in the early twentieth century, people didn't know this, and their descriptions of solid objects seem to go against it. Except that *we* describe solid objects the same way they did, even though we know contemporary atomic physics. Not to mention all the scientists who lived prior to Rutherford's epiphany held false beliefs about the nature of matter. So can we say that, since Darwin did not know that solid matter is mostly empty space, that everything he wrote about material objects (like finches) is thereby false? Of course not. The fact that the Bible refers to objects as they appear

---

18. Archer, *Encyclopedia of Bible Difficulties*, 93.

to us rather than in accordance with contemporary physics, geology, astronomy, or whatever, simply means, again, that it is using phenomenal terminology—just as we are when we do the same thing.

Other biblical passages that supposedly imply a flat earth are less direct. For example, the Bible says God "stretches out the heavens like a canopy, and spreads them out like a tent to live in" (Isa 40:22). This, some say, means that the universe is in the shape of a tent, with the earth representing the floor, implying it is flat. The obvious response is that it only says the sky above is *like* a canopy, *like* a tent. To suggest this must refer to its geometrical shape reads a lot of detail into the text that isn't there. Moreover, we can ask again whether this is describing the universe from some objective standpoint or is merely another case of phenomenal terminology, referring how the sky is above us, and comes down to the horizon around us, "like a tent." I see no reason to take this passage as an objective description of the universe's shape, or that it implies anything about the earth's. Similarly, Heb 8 describes the Jewish tabernacle as an image of the true tabernacle in heaven, and some of the same people suggested this means the universe is rectangular, like the tabernacle, with the earth representing the floor.[19] But this is an even worse argument: the tabernacle in heaven is not the physical universe but the spiritual realm transcending it. And even ignoring this, why take the analogy as having anything to do with shape?

Another claim: after his resurrection, Jesus ascended from the earth, and those nearby were told that he would return the same way—presumably by descending rather than ascending again (Acts 1:11). This is called the parousia or second coming, and it is believed by both Christians and Muslims. But we are also told that all the peoples of the world will see his return (Matt 24:30), even including those who killed Jesus (Rev 1:7) and people who have died (1 Thess 4:16). But if the earth is round, people on the opposite side of the globe will not be able to see Jesus' return, there being an entire planet between them and him. The only way for all people to witness the parousia is if the earth is flat.

Now I could point out these claims are hyperbolic, hyperbole being a common literary tool in all ancient Near East writing. I could also point out that by including people who are dead among the witnesses to the parousia, it's clearly talking about a supernatural phenomenon, as one might expect the end of the world to be, so I think this objection would

---

19. Hannam, *Globe*, 69–72.

be easily overcome. But I think we can pass over all that because modern technology has stepped in to resolve the problem. I'm pretty sure the second coming will be live-streamed.

Another passage involves the temptation of Jesus. Jesus was fasting in the wilderness (the Negev, southern Israel) for an extended period, and Satan tried to tempt him. One attempt to tempt was done by taking Jesus to the top of a high mountain and showing him all the kingdoms of the world and their splendor. Satan then says he'll give them all to Jesus if he worships him instead of God. Jesus declines the kind offer (Matt 4:8–10; Luke 4:5–8). Some people argue that Satan didn't show Jesus a vision during his fast, he took him to the top of a mountain so that Jesus could see far enough to actually observe all the kingdoms of the world from that height, thus implying that they thought the earth was flat. But this seems entirely forced: Israelites understood very well that the world extended beyond what one could see, even from on top of a mountain— especially since the highest mountains in the Negev aren't even 3,500 feet (a little over 1,000 meters) above sea level. You can see only about 75–80 miles (120–130 km) from that height. Jesus wouldn't have been able to see his hometown of Nazareth from there, but I'm pretty sure he still knew it existed. The Israelites couldn't see across the Sinai Peninsula to Egypt, the land of their enslavement, and they knew of other lands even further away, like Ethiopia and Yemen. Plus, first-century Israel was part of the Roman Empire, and they didn't think you could see Rome from Israel, not to mention the empire's western territories like Spain at the other end of the Mediterranean.

So the claim that the Bible states or implies that the earth is flat doesn't hold water. In saying this I'm not suggesting that it says it's round either. I don't think the Bible says anything about the shape of the earth.

Christians

The last argument I'll look at is the claim that plenty of Christians in the Middle Ages said that the earth was flat, and they were very influential. This counters the first contention above that says we know the medievals thought the earth was round because they said it was round. This objection says the opposite.

It's true that there are a few flat-earthers in history. I don't find this too discomfiting, since there are a lot more flat-earthers alive today. I live in Portland, and I'm willing to bet there are a few within a ten-mile radius of where I'm sitting right now.[20] But let's go over those foolish, flawed flat-earthers.

1. Lactantius (third and fourth centuries AD) is often listed as denying the earth was round, although his real target was those dreaded antipodes. Basically, he argued that people see the sun, moon, and stars moving from one horizon to the other, and assume that they are going around the earth so that the heavens are spherical; from this they conclude that the earth itself is also a sphere; from this they conclude that the other side of the earth must have just as much land as this side has; and from this they conclude that the other side must be populated.[21] It's not clear at which point he withdraws his consent—presumably he wasn't denying that the sun, moon, and stars go from horizon to horizon—but it certainly *sounds* like he was arguing against antipodes by arguing against a round earth. At any rate, his "views eventually led to his works being condemned as heretical after his death,"[22] although not because of this. Many Renaissance humanists admired his beautiful Latin (he was a professional rhetorician before becoming a Christian), but not because of the content of what he wrote.[23]

2. Diodore of Tarsus (fourth century) was an influential theologian. Some critics say he claimed the earth is flat, but the writings where he supposedly did so are lost. However, we don't have any particular reason to think his critics were lying or trying to defame him, so we can attribute a flat earth to him. Diodore was the foremost representative of the Antioch school of interpretation during its middle period, which tended to interpret the Bible literally unless there were unavoidable reasons not to, and which was dealt its almost-but-not-quite death blow at the Council of Ephesus in 431. His literalism led him to derive a flat earth from some of the Bible passages mentioned above, such as that God "stretches out the heavens like a canopy, and spreads them out like a tent to live in" (Isa 40:22). So, assuming his views were rightly expressed by his critics (which I think we should), we can ascribe a flat earth to Diodore.[24]

3. Theodore of Mopsuestia (fourth and fifth centuries) was a student of Diodore, and so followed the Antioch school's methods of interpretation.

20. To be fair, most of them probably came here from California.
21. Lactantius, *Divine Institutes* 3.24.
22. Russell, *Inventing the Flat Earth*, 32.
23. Rutherford, "Lactantius Philosophicus?"
24. Chapman, "Council of Ephesus."

As with Diodore, he is accused of affirming a flat earth for much the same reasons, but the writings in question are lost. In the absence of any contrary reason, we can attribute belief in a flat earth to him.[25]

4. St. John Chrysostom (fourth and fifth centuries) was also a student of Diodore and a companion of Theodore. He followed Diodore in arguing that the heavens are shaped like the tabernacle. However, unlike Diodore and Theodore, John's texts survive. Taking the parallel between the Jewish tabernacle and the heavenly tabernacle (Heb 8) literally, he asks, "Where are they who say that the heaven whirls around? Where are they who declare that it is spherical? For both of these notions are overthrown here."[26] So he argued that the universe isn't round. Since John was working with the assumptions of the Antioch school, and since he was a follower of Diodore, it seems reasonable that he also derives the earth's shape from the universe's, making it akin to a tent floor. John never explicitly makes this point, but it sure sounds like he's assuming a flat earth.

5. Severian of Gabala (fourth and fifth centuries) was another Antiochene interpreter. After Chrysostom graciously invited him to deliver some sermons, Severian thanked his benefactor by trying to have him banished. He denies the heavens are spherical, affirming the heavens-as-tabernacle interpretation. He further suggests that when the sun sets, it does not pass under the earth but goes around the horizon to the north, hidden from our sight. This doesn't directly state that the earth is flat, but it's close enough to confidently include him in the club.[27]

6. Cosmas Indicopleustes (sixth century) is the main person brought up as an example of medieval flat-earthery. Timing-wise he lived after the fall of Rome, but as he was in the Eastern Church, he belongs to late antiquity rather than the Middle Ages. Cosmas was a former merchant who gave extended descriptions of India and Sri Lanka, but he emphatically affirmed a flat earth as well, being heavily influenced by the Antiochenes, defending it extensively from the biblical passages already discussed, then drawing further conclusions. In this sense, then, he is an excellent example of a flat-earther since it informed much of his understanding of the Bible.

---

25. Baur, "Theodore of Mopsuestia."

26. "Homily 14," in Chrysostom, *Homilies*, 14:433. I don't want to guess what Chrysostom is suggesting when he challenges the idea that the heavens whirl around.

27. "Homily 3," in Severian, *Homilies on Creation and Fall*, 44; Baur, "Severian."

A few years after Cosmas published his book, *Christian Topography*, the "tabernacle universe" image was shot down by John Philoponus.[28] After that century, there are no positive references to it among the Nestorians[29] or in Eastern Christianity. No doubt Cosmas had continuing influence in some monastic communities as there are three surviving copies of his manuscript from the ninth to twelfth centuries.[30] But despite numerous claims to the contrary, Cosmas's overall influence on Eastern Christianity was minimal, and he had no influence on Western Christianity and Western civilization. Perhaps the best way to demonstrate this is by putting him in the context of the translation movement.

Starting in the twelfth century and continuing with fits and starts for another few centuries, many of the Greek writings that had been lost to the West were rediscovered, originally via the Islamic world's translations and commentaries of them. This led to a flurry of translating the ancient Greek texts into Latin (often via Arabic) to make them accessible to Western Europeans.[31] One of the last original translations was made by Copernicus in 1509, when he translated some shorter writings of Theophylactus Simocatta from Greek into Latin.[32] This was the first such translation published in Poland and so held some importance in that regard, but the text Copernicus chose was an insignificant one. The reason he selected Theophylactus is that all the good stuff had already been translated, much of it more than once. To make an original translation in 1509, Copernicus had to settle for the dregs.

So when was Cosmas translated? Short answer: in the early eighteenth century. Long answer: some manuscript collectors found Cosmas's work and translated a few excerpts as a curio in the late seventeenth century. It spurred some interest because of how bizarre it was, so the collectors translated the entire work into Latin and published it in 1706. This was long after the translation movement had ended, and nearly two hundred years after Copernicus translated one of the last untranslated works. So Cosmas's "very first introduction into western Europe" was

---

28. Hannam, *Globe*, 175.

29. Nestorianism has traditionally been considered a Christian heresy, but it is based on such fine points of doctrine that I've never been able to generate any condemnatory feelings for it.

30. Hannam, *Globe*, 176–77.

31. Lindberg, "Transmission of Greek and Arabic Learning."

32. Danielson, *First Copernican*, 50–51.

in the eighteenth century. "He had absolutely no influence on medieval western thought."[33]

7. Isidore of Seville (sixth and seventh centuries) used to be held up as an example of a flat-earther, but it's become a vexed issue. He was an encyclopedist, bringing together a bunch of information from diverse sources without trying to harmonize it, and it was sometimes unclear how invested he was in the claims he was relating. On the one hand, he called the earth a circle (*orbis terrae*) rather than a globe (*globus*) and portrayed the Arctic and Antarctic as touching each other. This would strongly suggest that he thought the earth was flat. On the other hand, he accepted Aristotle's cosmology of a spherical universe with a series of concentric spheres with the earth at the center, and that the spheres were equidistant from the earth on all sides. This would strongly suggest that he thought the earth was a globe—but only if he was consistent, which he wasn't. Historians used to consider Isidore a flat-earther, then changed their minds about it, and now some are changing their minds back again.[34]

So: we have three people who affirmed that the earth is flat (Diodore, Theodore, and Cosmas), three more who affirmed corollaries of it and so probably believed it's flat (Lactantius, Chrysostom, and Severian), and one more who may have believed it, but was radically inconsistent in his writings (Isidore). That makes a grand total of *seven* Christians who apparently believed that the earth is flat. Of these, four were followers of a specific school of interpretation that was quickly abandoned, and the latest two date to late antiquity and the very early Middle Ages, respectively.

Of course, these are *writers*. They would have had their supporters who never wrote down anything. However, these seven writers who affirmed or implied a flat earth are dwarfed by the number of Christian writers who affirmed or implied a *spherical* earth. If we can infer that the handful of flat-earth writers had their share of followers, then we can equally infer that the *much* more numerous spherical-earth writers had their share of followers too, and this leads to the conclusion that a spherical earth was the nearly universal view in antiquity and the Middle Ages. So the idea that flat-earthism was widely believed in ancient and medieval Christianity or Western civilization is simply false. "The truth

---

33. Russell, *Inventing the Flat Earth*, 35.
34. Isidore, *Etymologies* 14.2.1; *De Natura Rerum* 10.3; Hannam, *Globe*, 117, 207–11.

is that, after 800 AD, we don't know of anyone with a modicum of literacy who was unaware the Earth is spherical."[35]

With this, my original three points still stand. Why do we think the ancients and medievals thought the earth was round?

1. They said it was round.
2. They portrayed it as round.
3. Their cosmology required it to be round.

III. THE FLAT EARTH MYTH

So it turns out, there is a Flat Earth Myth, it's just not the myth that the earth is flat. It's a meta-level above that: the Flat Earth Myth is the myth that people before Columbus believed the earth was flat. This is a myth because people before Columbus did *not* believe that. But naturally this raises the question of how this urban legend got started in the first place.

"History" in scare quotes

The guilty party is Washington Irving, sometimes called America's Charles Dickens. His work of historical fiction, *History of the Life and Voyages of Christopher Columbus* (published in 1828) is apparently the earliest suggestion—at least as-yet discovered—that Columbus was trying to prove the earth is round.[36] Irving made Columbus's opponents the priests, inquisitors, and other members of the church hierarchy who took his claim as a challenge to their authority and the truth of Christianity. He also made it clear how dangerous it was to get on their bad side, a theme inspired by the trials of Galileo that continues to survive in the alleged conflict between science and religion.[37]

Anyway, Irving's "history" of Columbus permeated the culture, so that by the mid-nineteenth century most people in English-speaking countries had uncritically accepted it. There were a few reasons for this. First was the belief, common at the time, that we were heading towards a man-made utopia, and science was at the forefront of it. The advance of science made it

---

35. Hannam, *Globe*, 12.
36. Russell, *Inventing the Flat Earth*, 51–57.
37. See below, 102–6.

very easy to portray earlier ages as being so ignorant and unintelligent that they didn't even know such basics as that the earth is round.

Second, many people were taking the advance of science as justification to affirm *naturalism*. Roughly, this is the claim that the natural world is all that exists, entailing the rejection of any supernatural reality like God.[38] In order to make the connection between science and naturalism, they had to portray science and religion as enemies. This was the motivation behind finding examples of religion getting in the way of scientific advancement. The Flat Earth Myth was perfect for their purposes. The science in question was the very elementary discovery of the shape of the earth, the man proposing it was a freethinker, unafraid and unmoved by what people told him what he should believe, and his enemies were the religious authorities who threatened him with excommunication, torture, and death.

So the Flat Earth Myth isn't merely an urban legend: it's propaganda, misinformation, used to push the idea of a long trench warfare between science and religion. Of course, this doesn't mean that everyone who believed it (or believes it today) has any ulterior motives, but that was its original role. For that matter, those who originally used it to advance the cause of naturalism probably were not using it dishonestly; it just fit in so well with what they already believed that it made sense and perfectly illustrated their point—confirmation bias, again. That's why I say it was misinformation, not disinformation.

Isn't it ironic?

When I say that most people in English-speaking countries accepted the Flat Earth Myth in the nineteenth century, that includes Christians. And here it gets more interesting. Most Christians were embarrassed by the (false) claim that their faith traditionally taught that the earth is flat so they tried to ignore it or show that it wasn't an intrinsic part of Christianity. But in circumstances like these there are always some who take the road less traveled, although in many cases it's less traveled for a good reason. So when informed that Christianity traditionally taught a flat earth, some Christians responded by defending a flat earth. The most influential such author was Samuel Rowbotham, who published under the pseudonym

---

38. Giving it a precise definition is hard because the tendency is to say naturalism = anti-supernaturalism while supernaturalism = anti-naturalism. Thus, naturalism = anti-anti-naturalism, which is not very helpful.

"Parallax"—I presume because it sounds all sciency or something. He wrote plenty of tracts defending a flat earth, primarily by giving *scientific* reasons for it. He called his project "Zetetic Astronomy," and it not only included a flat earth but also the view that the earth is stationary.[39] One of Rowbotham's main arguments in defense of the flat earth was how the surfaces of some bodies of water were not curved as they should be if the earth were round. As proof he listed several lighthouses that could be seen from further away than they should if the surface of the earth was curved, but which made sense if it was flat.[40]

Rowbotham was not alone. Flat-earth societies sprang up all over the world in the nineteenth century to defend the "traditional," "Christian" view that the earth is flat, never mind the fact that the whole idea was only a few decades old. One intriguing case involves the Bedford Canal in Cambridgeshire, England.[41] Rowbotham argued for decades that over its six-mile length the canal was a flat surface rather than curved. Around 1870 one of his followers offered a bet to anyone who could refute Rowbotham, and Alfred Russel Wallace, codiscoverer of natural selection along with Charles Darwin, took him up on it. It ended with Wallace winning, the flat-earther threatening him in print repeatedly, and Wallace suing him. But the Bedford Canal continued to be a source for flat-earthery. Richard Proctor, a well-known astronomer of the time, had published a book where he referred to the Bedford Canal bet and argued that if, contrary to the events surrounding the bet, you could lie down with your eye close to the surface of the water and see something right above the surface six miles away, then "manifestly" the earth's sphericity would become problematic.[42] A flat-earther named Lady Elizabeth Blount took him up on it in 1904.

> She hired a photographer, Mr Clifton of Dallmeyer's, who in May 1904 went up to the Bedford Level, equipped with the firm's latest Photo-Telescopic Camera. The apparatus was set up at one end of the clear six-mile length, while at the other end Lady Blount and some scientific gentlemen hung a large, white, calico sheet over the Bedford bridge so that the bottom of it was near the water. . . . Mr Clifton, lying down near Welney bridge with his camera lens two feet above the water level, observed

---

39. Rowbotham, *Zetetic Astronomy*.
40. Garwood, *Flat Earth*, 36–78.
41. Garwood, *Flat Earth*, 79–117.
42. Proctor, *Myths and Marvels of Astronomy*, 277–79.

> by telescope the hanging of the sheet, and found that he could see the whole of it down to the bottom. This surprised him, for he was an orthodox globularist and round-earth theory said that over a distance of six miles the bottom of the sheet should be more than 20 feet below his line of sight. His photograph showed not only the entire sheet but its reflection in the water below.... The photograph was published in *The Earth* and other journals, and Lady Blount enjoyed her triumph. A long correspondence took place in the popular science magazine, *English Mechanic*, in which the orthodox tried to explain away the photograph as something to do with "refraction" or "mirages," while Lady Blount responded voluminously in prose, verse, songs and quotations from the Bible. She reminded readers of Richard Proctor's statement in one of his authoritative works on astronomy: "If with the eye a few inches above the surface of the Bedford Canal, an object close to the water, six miles distant from the observer can be seen, then manifestly there would be something wrong with the accepted theory." That test had now been made, and the accepted theory had come out of it badly. On the strength of it Lady Blount proclaimed that the earth's flatness had been scientifically proved.[43]

The author, writing in 1984, concludes that "there is obviously a need for further experiments at the Old Bedford Level,"[44] which implies that this test has never been refuted; or at least that the author was unaware of it. Similarly, many of Rowbotham's lighthouses have been destroyed, making refutation tricky. But obviously—*obviously*—this does not mean a flat earth is scientifically plausible. Rowbotham accurately copied the conditions of his lighthouses from a trustworthy sourcebook and found about thirty cases of lighthouses being visible from further away than "orthodox globularism" would dictate. But this is out of two thousand lighthouses listed in the book. The idea that these should overturn the sphericity of the earth is absurd and obscene. "The proper conclusion from the above facts is, that either there is a misprint in the book at these places, or that the localities where these lighthouses are situated possess some peculiarities which, if known, would account for these deviations."[45]

---

43. Michell, *Eccentric Lives and Peculiar Notions*, 29–30. This passage is how I discovered I am an "orthodox globularist."

44. Michell, *Eccentric Lives and Peculiar Notions*, 30.

45. Bresher, *Newtonian System of Astronomy*, 117–18. I found this reference via Schadewald, "Looking for Lighthouses," which is a good resource itself.

This flat-earthery was being defended by Christians in print into the twentieth century.[46] It had all but died out by around 1980, but recently flat-earth societies have begun springing up again, although they no longer seem to be affiliated with religion or Christianity. I suspect this is because the Internet allows like-minded people to congregate virtually and reaffirm their views without all that hullaballoo of interacting with people who disagree with them.

Myth preservation

Just as flat-earthism survives today, so does the Flat Earth Myth. When I was a kid, my sister Riki and I knew perfectly well that, long before Columbus, everyone knew the earth was round. I thought it was common knowledge. I thought that anyone who believed otherwise basically got their information from Bugs Bunny cartoons.[47] But plenty of actual historians repeat the myth so I can't lay the blame solely on popular culture. For example, in 2009, Jonathan Lyons used the Flat Earth Myth to contrast the alleged ignorance of medieval Christianity with the alleged erudition of medieval Islam. Ironically, his book is titled *The House of Wisdom*.[48]

As far as I'm concerned, the worst offender is Daniel Boorstin in his book *The Discoverers*. Chapter 14 is titled "The Flat Earth Returns," and is replete with talk of antipodes, the immense influence of Cosmas Indicopleustes, the danger one could get into by suggesting the heresy that the earth is round, all in order to portray medieval Christendom as monumentally foolish and antiscience. But it's all just crap. Later in chapter 20 ("Ptolemy Revived and Revised"), he writes that "no amount of theology would persuade a mariner that the rocks his ship foundered on were not real. The outlines of the seacoast, marked off by hard experience, could not be modified or ignored by what was written in Isidore of Seville or even in Saint Augustine."[49] So what theological claims would have any bearing *at all* on where rocks were located in the sea? Obviously there aren't any such claims. And where did Isidore or Augustine write about coastlines? Answer: they didn't. Boorstin made it up. This is a celebrated

---

46. Garwood, *Flat Earth*, 188–218.
47. "The earth is-a round! Like-a my head!" BONK! "The earth is-a *flat* like-a your head."
48. Lyons, *House of Wisdom*, 34–35, 46–47. See below, 183–85.
49. Boorstin, *Discoverers*, 148.

historian who was the librarian of Congress and senior historian of the Smithsonian Institute, as well as a distinguished professor at the University of Chicago and director of the National Museum of History and Technology. Yet his historical scholarship, at least on this point, is utterly unsalvageable. He should have known better. In fact I'm sure he *did* know better. But that didn't stop *The Discoverers* from becoming a bestseller.

Chapter 3

A Central Issue

THE FLAT EARTH MYTH is just freely invented, there's no truth behind it. Most of the other myths we'll be dealing with don't go this far. Instead, they're based on misunderstandings. Something true is twisted to mean something it didn't or exaggerated to suggest something it otherwise wouldn't. The present chapter is a case of twisting.

I. THE SCIENCE OF THE DAY

As we've already seen, the premodern view was that the universe is a series of concentric spheres with a spherical earth at the center. Additionally, premoderns believed there were four terrestrial elements, earth, air, fire, and water. Everything below the sphere of the moon was composed of some combination of these four. Water and earth made mud; earth and fire made stone; fire and water made whiskey; etc. But the spheres and everything above the sphere of the moon were thought to be composed of the *fifth* element: the aether or quintessence, the latter term being Latin for . . . wait for it . . . "fifth element." Unlike the lower four elements, the quintessence was not subject to degradation or decay. The quintessence completely filled the universe above the sphere of the moon. They did not think that a vacuum of empty space could exist.

This cosmology evolved over centuries but the basic idea goes back to the pre-Socratics. Aristotle developed it extensively in the fourth century BC.[1] In the second century AD Ptolemy expanded on it in multiple ways, making it much more detailed and mathematically rigorous. One important element was the use of epicycles: the stars didn't move in relation to each other (which is why they were called the *fixed* stars) but the planets (the wandering stars), as they moved across the sky over the year, would sometimes stop moving in relation to the stars, then start moving *backwards*, stop again, and then resume their forward motion. When this happens, it really means the earth is passing those planets in its orbit around the sun. But they didn't know that then, so Ptolemy, following the suggestions of some of his predecessors such as Hipparchus of Nicaea (second century BC), argued that each celestial sphere (a deferent) had another sphere embedded in it (an epicycle), and that sphere's planet moved around the epicycle sphere as the epicycle orbited the earth on its deferent sphere.[2]

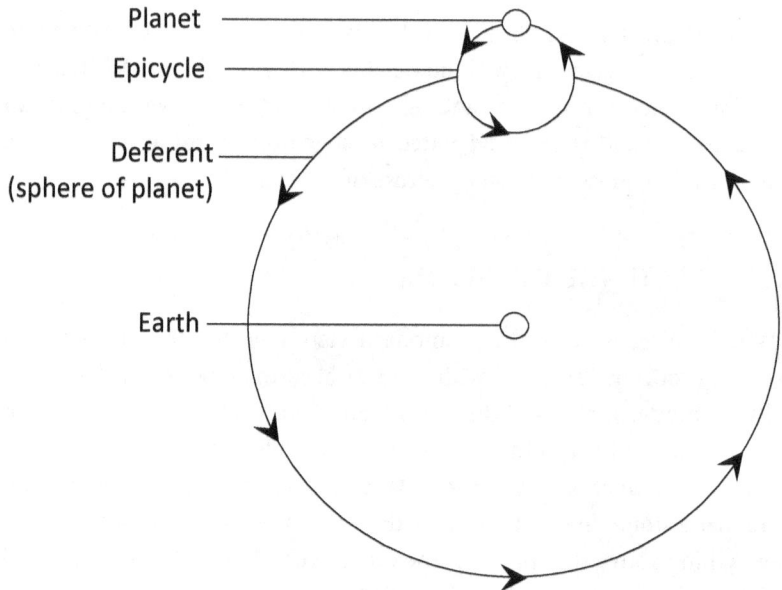

Ptolemy's cosmology was universal (or near enough) throughout Western civilization until the sixteenth century. Think about that: it was

---

1. Ring, *Beginning with the Pre-Socratics*, 138–39; Plato, *Timaeus* 48b; Aristotle, *On the Heavens* 1.2–3; *On Generation and Corruption* 2.3–5.

2. Lindberg, *Beginnings of Western Science*, 100–105; Murschel, "Structure and Function of Ptolemy's Physical Hypotheses."

established science for a millennium and a half. Darwinian evolution would have to be established for another twelve hundred years before it could be as entrenched as Ptolemaic cosmology was. It was accepted as given in all strata of society throughout antiquity and the Middle Ages, in technical and popular writings from Boethius to the *South England Legendary*.

> From the second to the sixteenth century, astronomy was a commentary on Ptolemy. No man ever wielded posthumously such a pervasive and long-lived authority in astronomy, and it is to be doubted that anyone ever will again. Ptolemy's work superseded the efforts of all his predecessors—surely one of the main reasons why so few of their works have survived—and it defined the astronomical problems for his successors, at least until the time of Tycho Brahe and Johannes Kepler.[3]

Obviously there's a great deal more that could be said here. Pierre Duhem's account of premodern cosmology takes up ten volumes.[4] But this is enough for our present purposes.[5]

II. GEOCENTRISM AND HELIOCENTRISM

The idea that the earth is at the center of the universe is called geocentrism. This means that the earth is neither moving through space nor spinning, it's just sitting there with the entire universe revolving around it. Why did they think this? Well, because we don't seem to be moving and everything else does. Pretty obvious. And since the same objects go past every day, they must all be moving *around* us. As noted, premoderns did not believe in empty space, all the vast distances of the cosmos were completely filled with the quintessence. The spheres were what the planets and stars were embedded in, and they were kept in motion by the Prime Mover, aka the Unmoved Mover, aka God, who keeps everything moving but is not moved by anything himself.[6]

In the third century BC Aristarchus of Samos made an alternative suggestion: perhaps the sun is at the center of the universe, not the earth.[7]

---

3. Van Helden, *Measuring the Universe*, 15.
4. Duhem, *Système du monde*.
5. Richard Garfinkle's science-fiction novel *Celestial Matters* is set in a world where Ptolemy's cosmology is actually correct. It's worth reading.
6. Aristotle, *Physics* 8.6; *Metaphysics* 12.6.
7. Hannam, *God's Philosophers*, 274.

This is called heliocentrism. Unfortunately, it was unable to overwhelm the observation that we don't seem to be moving. It wasn't until some scientists resurrected the old heliocentric theory and expanded upon it at the beginning of the modern era that it began to be defended again, first by Nicolaus Copernicus. During the rest of the sixteen century and into the seventeenth, Georg Joachim Rheticus, Thomas Digges, Diego de Zúñiga, Michael Maestlin, Thomas Harriot, Christopher Rothmann, Simon Stevin, Johannes Kepler, Giordano Bruno, Galileo Galilei, and possibly William Gilbert followed Copernicus and affirmed heliocentrism. Kepler's boss, Tycho Brahe, *kinda* did: he suggested that, apart from the moon, all the planets and stars orbit the sun—as the sun orbits the earth. However, since the earth was still at the center, Tycho's model was a modified geocentrism. All the above is on solid historical ground, none of it is urban legend.[8]

So, since people thought we were at the center of the universe, they must have also thought that, well, *we were the center of the universe*: we are the most important thing at the most important location in the entire cosmos. This equates *geocentrism* with *egocentrism*. To say something is at the center often has this meaning. If something is at the center of my attention, it means it is the thing I am most aware of; it is the most important thing to me, at least in that moment. We find this imagery in the Bible too (Ezek 48). So if we are at the center of the universe, we must be the most important thing in it. As Sagan puts it, "If the lights in the sky rise and set around us, isn't it evident that we're at the centre of the Universe? These celestial bodies . . . circle us like courtiers fawning on a king. Even if we had not already guessed, the most elementary examination of the heavens reveals that we are special."[9]

Why did people think they were so special? Because of religion. In Judaism and Christianity in particular, human beings are created in God's image (whatever that means) and are given authority over the other animals, which are *not* created in God's image. So we thought we were important because the Bible says so, and our examination of the heavens confirmed it by showing us that we're at the center of the universe. Thus when some scientists came around and started suggesting that maybe, just maybe, we aren't at the center of the universe, it was immediately perceived as an assault on the truth of Christianity and the authority of

---

8. Robert S. Westman, "Copernicans and the Churches," in Lindberg and Numbers, *God and Nature*, 76–113 (esp. 84–85); Hannam, *God's Philosophers*, 286.

9. Sagan, *Pale Blue Dot*, 20.

the church. It is a humiliation, a degradation of humanity to not be at the center. Sagan writes that this is part of "the series of Great Demotions, downlifting experiences, demonstrations of our apparent insignificance, wounds that science has, in its search for Galileo's facts, delivered to human pride."[10] The church was so offended by this dethronement that they imprisoned and tortured Galileo for affirming it.

III. SOME CRACKS IN THE ARMOR

So *some* of the latter half of that story is urban legend and some is true. Let's see what's what.

    1. As you should already know, this cosmology does not come from the Bible, from theology, from Judeo-Christianity, or from religion at all. It comes most notably from Aristotle and Ptolemy, and it was secular science (or proto-science). Ptolemy's cosmology was the science of the day for a millennium and a half. And Christians, including the church authorities, *accepted* the science of the day. In the thirteenth century Thomas Aquinas developed all of Aristotle's philosophy, including his cosmology, in a Christian direction.[11] When the church later sanctioned Aquinas's theology, this included Aquinas's sanction of Aristotle.

    But being *accepted* by Christianity doesn't change the fact that geocentrism does not come from Christianity but from secular thought. It was universal both within Christianity and without. Non-Christians accepted it too.

    2. The Galileo affair. I'll go into more detail about this in chapter 7. For now, I'll just point out that the issue is much more complicated than is usually presented. For one thing, Galileo had publicly mocked Pope Urban VIII, who, until then, had been a personal friend. Urban misused his power to get even with Galileo, and the heliocentric theory was the excuse he used. Another issue was that Galileo claimed that the Catholic clergy had erred in their interpretation of Scripture by accepting geocentrism. In Catholicism, laity (like Galileo) do not get to decide how to understand the Bible, only the clergy does. This was one of the main points of controversy in the Protestant Reformation. Galileo—unintentionally, I assume—had taken a Protestant stance as a Catholic layman less than a century after the start of the Reformation.

    10. Sagan, *Pale Blue Dot*, 20.
    11. Doig, "Aquinas and Aristotle."

3. The Galileo affair notwithstanding, the most vociferous denunciations of heliocentrism were not made by the church but by the scientific establishment of the time. They had, after all, devoted much of their scholarship towards expounding upon Ptolemy's geocentric cosmology, and were then being told that their paradigm was wrong and their lives' work worthless. I'm sure you'd be a little miffed too. Heliocentrism was seen as an ancient idea that had been decisively rebutted. Copernicus was a canon in the church and was encouraged by his fellow churchmen to publish his work *On the Revolutions of the Heavenly Spheres*, which he had finished by about 1532. However, he didn't arrange to publish it until right before he died because he feared negative reactions—from scientists, not the church.[12]

Some people claim that Martin Luther or John Calvin repudiated Copernicus for daring to contradict the Bible by suggesting the earth is not the center of the universe, but this is another urban legend. Luther apparently made an offhand comment that completely rewriting scientific orthodoxy, like heliocentrism did, is simply a cry for attention, but a) his objection had nothing to do with the Bible or theology and b) it was just an impromptu remark and didn't reflect some deep antipathy or anything. And there's no evidence that Calvin ever said anything against, or for, or *about* Copernicus or heliocentrism. He probably never heard about it.[13] The other early heliocentrists did not suffer any negative effects from the religious authorities as far as we can tell for their newfangled cosmologies, nor did Tycho for his modified geocentrism. Zúñiga later repudiated it (on Aristotelian, not biblical, grounds), and it's possible this was done under duress, but we don't know. Rheticus was blacklisted from academia, but not because he advocated heliocentrism. It was because he was accused (rightly or wrongly) of homosexual rape.[14] It's not until Bruno and Galileo that the Catholic Church got involved. They will be discussed in chapters 6 and 7.

4. This next one is an important point: *the scientific evidence at the time did not support heliocentrism*. If you were following the evidence wherever it led, you would have been a geocentrist. Often, when a new system replaces an old one, the old system explains almost everything, but there are hanging threads that are resistant to the system; plus, there

---

12. Hannam, *God's Philosophers*, 270–74.

13. Westman, "Copernicans and the Churches," in Lindberg and Numbers, *God and Nature*, 82.

14. Danielson, *First Copernican*, 143–50.

are some other threads that are explicable within the system, but only in very contrived, ad hoc ways. The new system explains these threads beautifully, but just by virtue of being new, it hasn't been wrung through the wringer enough to give it the same explanatory scope as the old system. Even in the second half of the seventeenth century, after Galileo's lifetime and *long* after Copernicus's, a stronger scientific case could be made for the earth being at the center rather than the sun.[15]

IV. THE MYTH OF DETHRONEMENT

All these issues are interesting and important in their own rights, but the most important one is the false equation of *geocentrism* with *egocentrism*. Geocentrism uses "center" literally: it says the earth is at the literal, physical, cosmographical center of the universe. Egocentrism uses "center" metaphorically: I am the most important thing in the universe. We can perceive things only from the perspective of our consciousness looking out at the world, making one's own mind the center. So, supposedly, we can't help but be the most important thing to ourselves. From this, it's an easy jump to anthropocentrism, where *human beings* are the most important thing, and from there to geocentrism, according to which the earth, being the standpoint from which human beings look out to the universe, is the most important place.[16]

The problem with this is that it is the *direct opposite* of the truth. In medieval cosmology, the center of the universe was the *least* privileged, *least* important, *least* valuable place to be. In a spherical cosmos, any movement towards the center is downward movement. The closer you get to the center, the further down you are. The earth wasn't at the center so much as it was at the *bottom* of the universe. Remember, they thought the sphere of fixed stars was the boundary of the physical universe, and God (the Prime Mover) was keeping all the spheres in motion from the outside. Well, what's the place within a sphere furthest removed from what's outside it? The center. They thought the center of the universe was the place furthest removed from God.[17]

---

15. Graney, *Setting Aside All Authority*.

16. Ted Chiang's excellent short story "Omphalos" is based on this assumption.

17. Although this bumps up against the doctrine of divine omnipresence: God is not located anywhere in the universe but is present everywhere within it. See Nicholas of Cusa, *On Learned Ignorance* 2.11. There were always tensions like this between Ptolemaic cosmology and the Abrahamic religions.

Also remember, the center was where the heavy elements fell, while everything above the sphere of the moon was composed of the quintessence or aether.

> Whatever was purest and most limpid (*liquidissimum*) rose to the highest place and was called aether. That which had less purity and some small degree of weight became air and sank to the second level. That which had still some fluidity but was gross (*corpulentum*) enough to offer tactual resistance, was gathered together into the stream of water. Finally, out of the whole tumult of matter all that was irreclaimable (*vastum*) was scraped off and cleansed from the (other) elements (*ex defaecatis abrasum elementis*), and sank down and settled at the lowest point, plunged in binding and unending cold. Earth is in fact the "offscourings of creation," the cosmic dustbin.[18]

The quintessence was not subject to decay and degradation, only the elements of the sublunar world were. And this was not just a physical truth but a *spiritual* truth. Above the sphere of the moon there was no spiritual or moral degradation, only the earth had sin and spiritual decay. The above quote understates it: the earth wasn't the universe's dustbin, it was the universe's toilet.

Yet, if the earth was the least important and least valuable place, why did the rest of the universe move in relation to it? Wasn't the center the point that defined the circular motion of the spheres? But in Aristotelian physics, motion is a *good* thing. The reason the earth didn't move is that it wasn't good enough to move. To be sure, there were some rumblings about this: in the fourteenth century, John Buridan and Nicole Oresme, while not challenging geocentrism itself, considered the possibility that the apparent motion of the skies was actually caused by the earth spinning in place. However, the idea was already ... *in circulation* ... as Francis of Meyronnes said it was being defended in the early part of that century before Buridan and Oresme. Ultimately, they determined that the scientific evidence was inconclusive (which, at the time, was true).[19] Regardless, since motion was good and circular motion was the most perfect type of motion, the center wasn't what defined the motions of the heavenly objects, the motions of the heavenly objects are what defined the center.

---

18. C. S. Lewis, *Discarded Image*, 62–63.
19. Hannam, *God's Philosophers*, 185–87; Copleston, *History of Philosophy*, 3:161–62.

V. INFERNOCENTRISM

For those still in doubt, there is an easy way to demonstrate the fact that the medievals believed the center of the universe to be the most insignificant and detestable place. The surface of the earth was not itself at the center: it was very *close* to it, but it wasn't quite there. To get to the center of the universe you'd have to dig, you'd have to go to the center of the *earth*. And what did they think was at the center of the earth? *Hell.* Hell was the center around which the rest of the universe orbited. Was hell a place of esteem, privilege, or value? Of course not, it was the outer darkness where one was finally and forever removed from God's presence. In fact, according to Dante at least, Satan was at the exact center of hell, thus the center of the earth, thus the center of the universe. This is why Arthur Lovejoy, the founder of the History of Ideas movement a hundred years ago, wrote that medieval cosmology should be called diabolocentrism rather than geocentrism. We don't have to be that exact, however, so I think we can settle for calling it infernocentrism.[20]

So they believed the earth was at the physical center of the universe but not at the metaphorical center, in the sense of being the most important place. What was at the metaphorical center? That should be easy: *God.* God, who is *outside* the furthest sphere, is the metaphorical center, the most important thing, in all existence. Thus, while ancient and medieval Christianity is not egocentric or anthropocentric, it is emphatically theocentric and Christocentric (Col 1:15–16). Naturally, this raises further questions. The literal center *is* often used to refer to the metaphorical center. That's why it works as a metaphor. So what is the reason for the reversal in this case, so that the literal center is the opposite of the metaphorical center?

> Because as Dante was to say more clearly than anyone else, the spatial order is the opposite of the spiritual, and the material cosmos mirrors, hence reverses, the reality, so that what is truly the rim seems to us the hub. . . . We watch "the spectacle of the celestial dance" from its outskirts. Our highest privilege is to imitate it in such measure as we can. The medieval Model is, if we may use the word, anthropo-peripheral. We are creatures of the Margin.[21]

---

20. Lovejoy, *Great Chain of Being*, 102. If we wanted to be *more* exact, we could say Satan's pancreas is at the exact center, making the medieval cosmos pancreatocentric, and I've suddenly lost all interest.

21. C. S. Lewis, *Discarded Image*, 58. I love the term "anthropo-peripheral" almost as much as "diabolocentric."

Hell and degradation were down, i.e., toward the center. Heaven and ennoblement, on the other hand, were up, i.e., *away* from the center. Christ *ascended* into heaven. Heaven is where there is no more sin. "Upward is the direction of improvement and rising importance (within Christianity, for example, Heaven is up; Christ rises from death and into Heaven; the spirits of the devout are exalted—literally, 'lifted high'—and so on). By contrast, downward, toward the center, is the direction of deterioration, corruption, and the grave."[22]

VI. THE REACTION OF THE CHURCH

I've already pointed out that most of the disapproval aimed at heliocentrism, at least at first (that's an important qualification), came from other scientists rather than the church. Granted, most scientists in Europe were Christians, and some of them, like Copernicus, were in the church hierarchy. But their objections to heliocentrism were not biblical or theological, at least for the most part (that's another important qualification); they were Aristotelian and scientific.

A direct theological objection to heliocentrism would have focused on the denigration of the sun by putting it in such a wretched place. The sun was the source of life and light and functioned as a picture and example of God's providence. To depreciate and deprecate it by putting it at the center could be taken as advocating contempt for God's love and care for us. Copernicus and Rheticus tried to rehabilitate the center, saying it wasn't necessarily the worst place, comparing it to a king sitting on a throne (which is probably where Sagan got the idea).[23] While this attempted to forestall the objection, it didn't work: as Giovanni Tolosani, a fellow sixteenth-century astronomer, wrote, "Copernicus puts the indestructible sun in a place subject to destruction."[24]

The flip side of this is that, by putting the earth among the stars, they were putting it in the perfect, incorruptible heavens. This could imply that the earth and its denizens were not so bad that they needed atonement, salvation, and the church's mediation of God's grace. I'm unaware of this

---

22. Danielson, "Great Copernican Cliché," 1035.

23. Dennis R. Danielson, "Myth 6," in Numbers, *Galileo Goes to Jail*, 50–58 (esp. 54–55); Danielson, *First Copernican*, 77–78, 82; Sagan, *Pale Blue Dot*, 20.

24. Danielson, "Myth 6," in Numbers, *Galileo Goes to Jail*, 54.

being made as an *objection* per se, but the early heliocentrists indicated that their theory elevated the status of the earth and humanity.[25]

The Bible and geocentrism

The complete predominance of Ptolemaic cosmology had led some theologians over the centuries to try to retcon geocentrism back into the Bible. They found some passages that refer to the earth as not moving: "Indeed, the world is firmly established, it will not be moved" (1 Chr 16:30; Pss 93:1; 96:10); "He established the earth upon its foundations, so that it will not totter forever and ever" (Ps 104:5). These verses *could* be taken that way, as if they were speaking from an objective unmoving standpoint out in space looking down upon the earth and defining the earth's lack of motion by virtue of its correlation to that objective point's lack of motion. Or perhaps they thought it was describing it from God's perspective outside the universe, independent of space and hence motion (which would make more sense). But it's more plausible that the point of view in question is just the surface of the earth, where all the people reading these texts live. From *that* standpoint, yeah, the earth is firmly established and will not be moved. Sure, there are the occasional disasters that upend things—earthquakes, volcanic eruptions, Shai-Hulud—but this doesn't affect the general truth that the earth is not usually tottering back and forth when you're out on your morning constitutional.

Other passages attribute the movement of the heavenly objects across the sky to their movement rather than the earth's: "The sun rises and the sun sets; and hastening to its place it rises there again" (Eccl 1:5); and "in them [the heavens] He has placed a tent for the sun, which is as a bridegroom coming out of his chamber; it rejoices as a strong man to run his course. Its rising is from one end of the heavens, and its circuit to the other end of them" (Ps 19:4–6). But we use the terms "sunrise" and "sunset" despite knowing that these phenomena are caused by the earth's motion rather than the sun's. From the earth's surface, such terms are completely appropriate because the "sky" simply indicates what we see when we look up. These are more examples of phenomenal terminology which refer to how things *appear* without making any claim whether they would appear differently from another standpoint. So taking any of these Bible passages as affirming geocentrism requires a lot of assumptions that are not given.

25. Danielson, "Myth 6," in Numbers, *Galileo Goes to Jail*, 55–56.

The best case, supposedly, is Josh 10:12–14. The Israelites were in battle against the Amorites, and God made the sun stand still to extend the daylight, allowing Israel to defeat their enemies. The whole idea suggests that the sun's motion was halted, and thus that it was the sun that moves, not the earth (although it is kind of weird that the best example of the sun's movement comes from a passage saying it stood still).

Now, of course, if the earth stopped spinning on its axis so that the sun appeared to stand still, all life on earth would be destroyed. But in a geocentric universe, the sun stopping in the sky would destroy all life on earth too (the earth would burn up, in fact). So whatever problems one may have about the plausibility of this account are not resolved by putting the earth at the center. But ignoring this, the only support it gives to the geocentric side is that it says the sun stopped, just as other passages say that the sun moves, rises, and sets. This is simply phenomenal language again. From the perspective of someone on earth, that's what it would have looked like. So no, this passage doesn't provide any more support for geocentrism than the terms sunrise and sunset do.[26]

Fictionalism

One issue throughout philosophical and scientific history has been whether we should take some models, hypotheses, or theories as merely instrumental (saving the appearances or phenomena) or as literal, physical accounts. Regarding heliocentrism, the question was whether it was an actual physical description or more like a mathematical model that explained things without asserting its objective truth. Today, this sort of idea is often called fictionalism. There are fictions throughout academia where ideas are used to explain a phenomenon or set of phenomena, with the full understanding that the idea is not literally true. There are fictions in science (ideal gases), mathematics (points), philosophy (possible worlds), political theory (social contracts), and much else besides. Particle physics is rife with them. Applied to the issue at hand, some

---

26. This is one of those Bible passages that some people think is too crazy to take seriously. If it was a historical event (and an objective one, not a vision or something like that), all we could say for sure about it is that God supernaturally kept the battlefield illuminated beyond the point where it normally would have been. This is pretty tame as far as miracles go. See ch. 13. Alternatively, the text may not even claim that the sun stopped moving. On this interpretation, see John Walton, "Biblical Credibility and Joshua 10: What Does the Text Really Claim?," https://biologos.org/articles/biblical-credibility-and-joshua-10-what-does-the-text-really-claim.

people wondered whether heliocentrism was a form of fictionalism that explained those phenomena resistant to the geocentric model but which shouldn't be taken as an actual account of reality or whether it was more than just instrumental.

Part of the problem was whether physics or mathematics had priority over the other. Traditionally, it had been physics, but the heliocentrists were claiming that a mathematical model, like heliocentrism, could also be physically correct. This was an astonishing claim as it contradicted the way science had been done for centuries,[27] which was one of the reasons so many scientists were opposed to it.

Egocentrists

So how did this equation between geocentrism and egocentrism (or anthropocentrism) arise? As mentioned, Copernicus and Rheticus tried to rehabilitate the center by comparing it to a king sitting on a throne, but it didn't catch on until the decades following Galileo's death when some French satirists suggested the popular storyline that decentralizing the earth was a sort of dethroning or demotion.[28] Until then it was seen unequivocally as a huge *promotion*. Galileo himself portrayed heliocentrism as making the earth part of the "dance of the stars" rather than "the sump where the universe's filth and ephemera collect."[29] So it wasn't a dethronement unless you're talking about the porcelain throne.

This raises the question whether there were people before Copernicus who thought being at the center of the universe was a good thing. One suggestion is Seneca (first century AD), who wrote, "That you may understand how she [Nature] wished us, not merely to behold her, but to gaze upon her, see the position in which she has placed us. She has set us in the center of her creation, and has granted us a view that sweeps the universe."[30] But here, Nature is the subject and wants an audience. The center is not a place of privilege *except* in that it allows us to see Nature's glory to the maximal degree. The rest of the universe deserves to be observed; the earth does not. The center is the box seats that allow the fans

---

27. Westman, "Copernicans and the Churches," in Lindberg and Numbers, *God and Nature*, 78–81.

28. Danielson, "Copernicus and the Tale of the Pale Blue Dot."

29. Galilei, *Sidereus Nuncius*, 57.

30. Seneca, *De Otio* 5. Calcidius says something similar; see C. S. Lewis, *Discarded Image*, 55.

to have the fullest view of the game. But the field and participants in the game are more important than the fans and their seating arrangements, since they're what the game is *about*.

A better example is Saadia Gaon, a Jewish rabbi who lived in the Muslim world during the tenth century. He argued that nature tends to put the most important things at the center: seeds for fruit, yolks for eggs, hearts for animals and people, so it stands to reason that if the earth is at the center of the universe, it's the most important part of the universe. This is just to affirm the easy correspondence between the literal center and the metaphorical center, despite the explicit reasons to the contrary. Besides which, at best, this would mean that the *earth* is important, but why should human beings be more important than any other objects on earth? They're all equally at the center, after all. Rocks, slugs, and fungi have as strong a claim as being the most important thing on Saadia's grounds. But it's all academic anyway since he didn't influence anyone on this point as far as we can tell.[31]

VII. NEOCENTRISM

So there isn't any more reason to think medieval people thought the earth was the most important place by virtue of being at the center of the universe than there is to think medieval people believed the earth was flat before Columbus. But just as flat-earth societies started popping up in the nineteenth century, so today there are several Christian ministries dedicated to defending geocentrism, and unfortunately (and unlike contemporary flat-earth societies) they are still primarily associated with Christianity.

The first one I discovered was beyond disappointing. In my mid-twenties I became a Christian and in particular a Protestant (although I have plenty of Catholic sympathies). Because it happened later in life, I don't have as much of an emotional attachment to Protestant doctrines over Catholic doctrines that many people who have been raised in these traditions have. This leads me to being less secure in my side's position. While Catholics and Protestants agree on about 99 and 44/100 percent of their theologies, there are important points of contention, one of the most notable being the connection between faith and salvation. Protestants say we are saved by faith alone (*sola fide*) while Catholics say we have to have good works in addition to faith to secure our salvation. Much of this

---

31. Brague, "Geocentrism as a Humiliation for Man," 193–96.

theological hullabaloo is based on misunderstanding: the two sides were using different definitions of "faith" and "salvation," and when corrected a great deal of the controversy dissipates.[32]

Anyhoo, I discovered a book by a Protestant-turned-Catholic named Robert Sungenis that criticizes the Protestant position on faith and salvation and defends the Catholic one. It's a beautifully thick book with a beautifully provocative title: *Not by Faith Alone*. So I was planning to set aside a summer and pore over this book, along with *Iustitia Dei*, a book by a Protestant (Alister McGrath) that does something similar in the other direction, and see where I came out. Unfortunately, before I could do so, I discovered that Sungenis had published a multivolume work titled *Galileo Was Wrong, the Church Was Right*, which defends geocentrism from a scientific perspective (volumes 1 and 2) and a theological perspective (volume 3). So I'm afraid I just don't trust his judgment. I'm not associating the Catholic doctrine of justification with geocentrism or anything, I'm simply unable to take this particular author seriously because of what he's written in defense of geocentrism. I have no desire to read him anymore.

Another geocentric ministry is the Association for Biblical Astronomy, although they prefer the term "geocentricity" as it doesn't have all that historical baggage.[33] There are plenty of other websites like this, but fortunately there are also books and websites that rebut them scientifically and theologically.[34] Although most young-earth creationists roundly reject geocentrism, some young-earth ministries have published articles defending it.[35] The Institute for Creation Research has an article on their website defending geocentrism as something young-earth creationists should accept.[36] Another young-earth advocate tried to argue that our universe is the interior of a white hole (a black hole where the matter becomes so compressed that light is forced out via quantum tunneling). He suggests that if we were at the center of this white-hole universe, time would pass slower than it would on the outskirts. To be fair, he's only

---

32. As Pope John Paul II noted to a group of Lutheran bishops. See Kreeft and Tacelli, *Handbook of Christian Apologetics*, 320–21.

33. http://www.geocentricity.com.

34. Keating, *New Geocentrists*. See also the Geocentrism Debunked website at http://www.geocentrismdebunked.org.

35. Keating, *New Geocentrists*, 157–58. Young-earth creationism is the idea that the universe was created by God only a few thousand years ago.

36. Aardsma, "Geocentricity and Creation."

suggesting that the Milky Way Galaxy is at the center of the universe so the earth is only approximately at the center—galactocentrism rather than geocentrism. But why does he think this? He states it plainly:

> Most important, it is very encouraging to see evidence for the centrality of humans to the plan of God. It was a sin on *this* planet that subjected the entire universe to groaning and travailing (Romans 8:22). Ours is the planet where the Second Person of the Trinity took on the (human) nature of one of His creatures to redeem not only us, but also the entire cosmos (Romans 8:21). This knowledge that God gave minuscule mankind prime real estate in a vast cosmos astounds and awes us, as Psalm 8:3–4 says: "When I consider your heavens, the work of your fingers, the moon and the stars, which you have ordained; What is man, that you are mindful of him? and the son of man, that you visit him?"[37]

This is precisely the "geocentrism = egocentrism" equation that is the core of the Myth of Dethronement and that has been used to fuel the claim that science and religion are always butting heads.[38] And just like some Christians bought into the Flat Earth Myth and defended a flat earth to defend Christianity, so some people are buying into the Myth of Dethronement and defending geocentrism to defend Christianity. I would suggest to both camps that they shouldn't let their opponents dictate the terms of the debate—specifically when those terms are what the Bible says and what the Christian church has historically taught.

---

37. Humphreys, "Our Galaxy Is the Centre of the Universe, 'Quantized' Redshifts Show."

38. *Butting* heads, not *butt*heads (although that may be true too).

Chapter 4

Cosmos of Unusual Size

So premodern cosmology was characterized by a series of concentric spheres, with earth at the center, and with each successive sphere embedding the moon, sun, and planets, out to Saturn. Beyond that was the sphere of the fixed stars. This final sphere was the boundary of the physical universe. Yet this is smaller than our *solar system* since it doesn't take Uranus and Neptune into account. Actually, it's worse than this since they severely underestimated the distances. Ptolemy calculated the distance to the outer sphere to be 19,685 earth radii—about 80 million miles or 130 million kilometers—although he also states that this is the *minimal* distance.[1] That's less than the distance from the earth to the sun. But even if they had accurately assessed the distances to the other planets, we have learned that the universe is orders of magnitude larger than our solar system. Just watch one of those videos contrasting the size of the earth with the sun, with larger stars, with galaxies, and you'll get the idea.

So, yes, medieval people might have thought the center of the universe was the most degraded and valueless location, but they still might have thought that human beings were the most important thing because they were the biggest thing in a small universe, thus justifying the exalted view of humanity that religion posits. And our discovery of the unimaginable vastness of our cosmos takes this possibility away. We see this point in popular culture as well as academia. In *The Hitchhiker's Guide to the Galaxy* mythos there is a torture device called the Total Perspective

1. Goldstein, "Arabic Version of Ptolemy's Planetary Hypotheses," 7–11.

Vortex, which gives people a glimpse of the enormity of the universe and themselves in relation to it. I say "torture device" but really it's a complicated and cruel method of execution since no one can survive having a sense of proportion.[2] Monty Python's *The Meaning of Life* has a Universe Song that expresses similar sentiments. Or if the Universe Song from Monty Python is not family friendly enough for you, try the Universe Song from Animaniacs. But I'm a philosopher so as far as I'm concerned the best expression of this idea comes from the beginning of Friedrich Nietzsche's brilliant essay "On Truth and Lies in a Nonmoral Sense":

> Once upon a time, in some out of the way corner of that universe which is dispersed into numberless twinkling solar systems, there was a star upon which clever beasts invented knowing. That was the most arrogant and mendacious minute of "world history," but nevertheless, it was only a minute. After nature had drawn a few breaths, the star cooled and congealed, and the clever beasts had to die.[3]

As you can guess, though, this idea is based on a misconception.

I. THE BIG FISH IN A SMALL POND MYTH

Of course it's true that the Ptolemaic universe was small *in comparison* to how large we've discovered it to be. But that doesn't mean the premoderns thought the universe was small in an absolute sense. In fact, they thought the universe was larger than we can imagine and the earth was so small in comparison that, for mathematical purposes, it should be treated as a point of zero volume.[4] If you take an unimaginably large number, call it A, and multiply it by itself, you reach another number, B, which is not only unimaginably large, it's unimaginably *larger* than A. Then multiply B by itself and reach C. Keep squaring numbers all the way up to Z. Z is an unimaginably large number as well as being unimaginably larger than Y, which is unimaginably larger than X, and so on. So how much larger

---

2. Except Zaphod Beeblebrox, since he actually *is* the most important thing in the universe—at least that particular "universe" where he went into the Total Perspective Vortex.

3. I *love* this. How can anyone *not* love this? Nietzsche, "On Truth and Lies," 14. It was originally written in 1873 and published in German in 1896 as *Über Wahrheit und Lüge im außermoralischen Sinne*.

4. Van Helden, *Measuring the Universe*, 4–5.

is *Z* than *A*? Unimaginabilities upon unimaginabilities. In comparison to *Z*, *A* is so tiny as to be infinitely small.

But remember, we *started* from the premise that *A* is unimaginably large. Saying that all the numbers after it in our exercise make it as nothing does not take away from the fact that it is still unimaginably large. That there are larger numbers does not make *A* objectively small, only small *in contrast* with the larger ones. And when it comes to unimaginably large sizes or distances or masses or whatever, we are able to differentiate them from each other only mathematically, not imaginatively. That's what makes them unimaginable.[5]

So Ptolemy estimated the universe's radius as 80 million miles. Over the next 1,500 years or so, there were plenty of corrections, recalculations, and adjustments made, but for the most part they were in the same general ballpark. But these distances are unimaginably large. So they were starting with an unimaginably large universe, and the fact that we have discovered that it's unimaginably larger to an unimaginably large degree does not suggest they thought the universe was a small place. None of this, incidentally, reduces the incredible value of how contemporary science has discovered the vastness of the universe, it just repudiates using that discovery in the way described.

That the ancients and medievals thought the universe was unimaginably large and the earth a mathematical point in comparison is not a contentious claim. Aristotle wrote, "Observation of the stars also shows not only that the earth is spherical but that it is of no great size." Ptolemy wrote, "The earth has sensibly the ratio of a point to its distance from the sphere of the so-called fixed stars." Boethius wrote, "You have learned from astronomical proofs that the whole circle of our earth is but a point in comparison with the extent of the whole heavens; that is, if it is compared in size with the celestial sphere, it is judged to have no size at all."[6] This was an essential element of the cosmology of ancient Greece and Rome, which was accepted by ancient and medieval Christianity, and it was acknowledged by everyone across the board. "The spatial insignificance of Earth . . . [was] asserted by Christian philosophers, sung by Christian poets, and commented on by Christian moralists for some

---

5. C. S. Lewis, *Discarded Image*, 98–99.

6. Aristotle, *On the Heavens* 2.14 (297b31–33); Ptolemy, *Almagest* 1.6; Boethius, *Consolation of Philosophy* 2.7, 10–14.

fifteen centuries, without the slightest suspicion that it conflicted with their theology."[7]

More than this, though, is the fact that we are spatial beings and cannot imagine the absence of space (the same goes for time).

> Whatever space may really be, it is certain that our perceptions make it appear three dimensional; and to a three-dimensional space no boundaries are conceivable. By the very forms of our perceptions therefore we must feel as if we lived somewhere in infinite space: and whatever size the Earth happens to be, it must of course be very small in comparison with infinity.[8]

Psychologically, we are naturally inclined to think that the universe is so big that we are as nothing in comparison to it. So the very idea that the ancients and medievals would have thought that the earth is the biggest thing in a small universe, thus justifying the religious idea that human beings are important, collapses right out of the gate.

II. OBJECTIONS

A potential counterargument is that, even though they thought the distances between objects were vast, perhaps they thought earth was still the largest object in the universe, and important in that sense. But this is false: in the Ptolemaic system, only Mercury, Venus, and the moon were thought to be smaller than earth (as they are). Everything else was bigger. The smallest star was many times larger than the earth.[9] So our earth was one of the *smallest* objects in an unfathomably large universe. Besides which, all those vast distances weren't empty space since they didn't believe a vacuum could naturally occur. Instead, they thought beyond the sphere of the moon the universe was completely filled with the fifth element,[10] and this is what formed the concentric spheres. This means that each sphere was an object. The moon was smaller than the earth, but the *sphere* of the moon was much, much larger—and that's the *innermost* sphere. The sphere of the fixed stars was an absolutely enormous object and simply dwarfed the earth.

---

7. C. S. Lewis, *Miracles*, 53.
8. C. S. Lewis, *Miracles*, 53–54.
9. Lovejoy, *Great Chain of Being*, 100.
10. Boron. Ha! No.

Another potential objection is that in premodern literature, characters are sometimes taken outside the earth to the sphere of the moon, or even to the sphere of fixed stars. From this vantage point, they then look down upon the earth and see all kinds of details that would be impossible to see from a great distance. Doesn't this suggest that they did not actually conceive the distances to be very great? That turns out to be a big nope.

> The impossibility, under the supposed conditions, of such visual experiences is obvious to us because we have grown up from childhood under the influence of pictures that aimed at the maximum of illusion and strictly observed the laws of perspective. We are mistaken if we suppose that mere commonsense, without any such training, will enable men to see an imaginary scene, or even to see the world they are living in, as we all see it today. Medieval art was deficient in perspective, and poetry followed suit. Nature, for Chaucer, is all foreground; we never get a landscape. And neither poets nor artists were much interested in the strict illusionism of later periods. The relative size of objects in the visible arts is determined more by the emphasis the artist wishes to lay upon them than by their sizes in the real world or by their distance. Whatever details we are meant to see will be shown whether they would really be visible or not. I believe Dante would have been quite capable of knowing that he could not have seen Asia and Cadiz from the *stellatum* and nevertheless putting them in. Centuries later Milton makes Raphael look down from the gate of Heaven, that is, from a point outside the whole sidereal universe—"distance inexpressible By Numbers that have name" (VIII, 113)—and see not only Earth, not only continents on Earth, not only Eden, but cedar trees (V, 257–61).[11]

III. A FEW MORE ISSUES

God's vastness

Another point to make here is that the immensity of the universe was often taken as a reflection of God's greatness. In most languages, the word for "great" starts off meaning really, really big because we subconsciously

---

11. C. S. Lewis, *Discarded Image*, 101, referencing John Milton's seventeenth-century epic poem *Paradise Lost*.

associate size with importance (I blame Texas for this). Looking at the night sky and recognizing our own insignificance in light of the enormity of the cosmos is one of the most common triggers of religious beliefs, since it forces us to recognize that there is something vastly more important than ourselves. When drawn out, the idea is that the immense size of the universe, when compared with the size of the earth and humanity, makes us think of the prime reality, next to which we are as nothing. As big as the universe is, God is, if I may put it this way, bigger. The universe reflects God's immensity, his greatness. For the ancients and medievals, this experience was magnified with a more specific belief of exactly how big the universe was, and the size in question was larger than they (or we) could imagine.

This is also true of the universe's age. As old as the universe is, God is even older. A cosmos billions of years old reflects God's eternity since we cannot imagine billions of *anything* and just see it as "more than can be counted." But God is even more ancient than that. In fact, the Bible specifically says that God's eternity is seen through his creation: "Ever since the creation of the world his *eternal power* and divine nature, invisible though they are, have been understood and seen *through the things he has made*" (Rom 1:20). This is more than a little ironic, since the young-earth creationists mentioned above believe that the Bible obligates them to affirm that the universe is only a few thousand years old. But a few thousand years are *imaginable* and would not reflect God's eternity.

Infinitude

During the Renaissance, Nicholas of Cusa, Thomas Digges, Giordano Bruno, and others argued for an unbounded universe: if God is infinite and the universe's vastness reflects God, then the universe must be infinite as well. More recently, in 1957, Alexandre Koyré published *From the Closed World to the Infinite Universe*, which went over the effects of moving away from the premodern cosmology to the modern, and specifically the change from a finite cosmos to an infinite one. His point was that an infinite universe is not just bigger: it's a different kind of thing than a finite universe, even one that's unimaginably large.

> Let us not forget, moreover, that, by comparison with the infinite, the world of Copernicus is by no means greater than that of mediaeval astronomy; they are both as nothing, because *inter*

finitum et infinitum non est proportio [between the finite and infinite there is no proportion]. We do not approach the infinite universe by increasing the dimension of our world. We may make it as large as we want: that does not bring us any nearer to it.[12]

I think this is absolutely correct, they're totally different beasts. It seems to me that an infinite universe would produce a sort of *horror vacui*: a sense of agoraphobia or kenophobia (fear of empty space) or apeirophobia (fear of infinity) since the mind cannot rest in the infinite. There is no absolute standard for reference, everything is always smaller than something else. This sounds like Pascal's famous line, "The silence of the infinite spaces terrifies me." The same terror of the infinite also nearly drove Martin Buber to suicide.[13] None of these phobias would have touched the premoderns as they believed the universe to be finite, filled with the quintessence, and deeply structured. But how would any of this amount to an argument against God's existence or religion or Christianity?

Well, it doesn't, and Koyré doesn't suggest otherwise. But perhaps there is a subtle point we could make here: in an infinite universe we are small, but so is everything else. The galaxies are small compared with something larger, etc. There's no stopping point. But by the same token, everything is *large* compared to something else. Compared to the microscopic world, we're huge, and compared to the atomic world, the microscopic world is huge. This, as I say, is what may produce this inability of the mind to rest, since there's no absolute standard of reference. But the corollary of this is that a *finite* universe can make our smallness more powerfully felt, since there *would be* an absolute standard of reference. We're larger than some things, sure, but we're smaller—much, *much* smaller—than the largest thing (viz., the universe itself).

So what would this mean for Christianity? Just that an unimaginably large but finite universe might portray God's vastness better than an infinite universe would,[14] contra Nicholas, Digges, and Bruno. But this may be splitting hairs, and in any event I don't think an infinite universe can be developed into an argument against theism or Christianity. At most we could say that it takes away one of our pointers to God, if that, but that doesn't amount to much of an objection.

---

12. Koyré, *From the Closed World to the Infinite Universe*, 34.
13. Pascal, *Pensées*, 64; Friedman, *Martin Buber's Life and Work*, 27, 96.
14. C. S. Lewis, *Discarded Image*, 99–100.

All of this is somewhat academic though, since big bang cosmology demonstrates that the universe is spatially, temporally, and materially finite. It is expanding because there is a finite amount of matter taking up a finite amount of space that came into being a finite amount of time ago. To be sure, these amounts are all insanely (*unimaginably*) large, but "between the finite and the infinite there is no proportion."[15] And at any rate, the overall point remains the same: the ancients and medievals did not base their religious belief that human beings have value on the idea that the earth is the biggest thing in a small universe.

Anthropic coincidence #1: Mass density

Two specifically scientific issues arise here from the Anthropic Principle. This refers to the conditions that must be met for a planet, solar system, galaxy, and universe to support life. It is a stunningly bad name because "anthropic" refers to humanity, not life in general. Scientists suck at naming things.[16] It should be called the biophilic principle, or maybe the carbophilic principle, but nobody asked me.

The first issue deals with the universe's mass density. This is the density of matter and energy, which sorta[17] correlates to the amount of matter created in the big bang. The point I want to make is that if the mass density were much different than what it is by a tiny amount, about one part in 10^{60}, then biological life could not exist at any place, at any time in the universe's history.[18] This is because the density of matter and energy that exploded into existence with the big bang dictated how gravity affected the universe's initial expansion from said explosion, matter being what gravity acts upon. This was one of the main motives for developing cosmic-inflation theory. Less matter means less of a gravitational influence, means the matter expands faster, means it goes too fast to clump together and form stars and galaxies, and ends up as an enormous cloud of hydrogen atoms with a smattering of helium and deuterium atoms mixed in. More matter means more of a gravitational influence, means

---

15. Koyré, *From the Closed World to the Infinite Universe*, 34.

16. In saying this I stand firmly in the Calvinist tradition. See Watterson, *Homicidal Psycho Jungle Cat*, 84.

17. Here I'm using Daniel Dennett's "*sorta* operator." If he gets to use it, so do I. See his *Intuition Pumps and Other Tools for Thinking*, 96–97.

18. Barrow and Tipler, *Anthropic Cosmological Principle*, 411; Krauss, "End of the Age Problem," 465.

the slower the matter expands, means it goes too slow and collapses on itself to create a giant black hole. In neither case would you get biology. In the vast majority of cases, you wouldn't even get *chemistry*. So in order for life to exist—including, you know, *us*—the universe has to have approximately the amount of matter it has, spread out at just the right density through space. Otherwise, there would be no stars, no planets, no life.

Now some people may be underwhelmed by this. Sure, this matter might have had some relevance to the universe's initial expansion billions of years ago, but there's no connection between distant pieces of matter and the human race's present existence. There are plenty of galaxies billions of light years away, which have billions of stars with billions of planets orbiting them. What does a particular rock on a moon orbiting one of those planets have to do with life on earth *now*? The absence of such a connection shows that humanity is irrelevant to the universe and so we are utterly insignificant.

In response, first, as I have been at pains to show, *the premoderns already believed this*. They thought the earth was smaller than almost everything else. The smallest star was many times larger. So if we picked a star at random, what did one mole of hydrogen (or quintessence, rather) of that particular star, two-thirds of the way from its surface to its center, have to do with humanity? *Nothing*. This is not a new discovery brought about by the advance of science. If this is an argument against Christianity or theism, it would have been just as effective two thousand years ago under the Aristotelian/Ptolemaic cosmology.

Second, so what? So what if all that matter has no relevance to us or we to it? Why should it? Why should it need to have relevance to humanity at all? God can appreciate everything he's created on its own terms. God delights in empty space, inert matter, living matter, conscious matter, and in whatever else may exceed this.

Besides which, what exactly is being asked here? That once objects are no longer relevant to our existence, God should snuff them out? Wouldn't that be to *remove* the evidence? So if he leaves it there once it's no longer relevant to us, its present irrelevance implies it never *had* any relevance to us, so it never had any *purpose*, and so God does not exist. If he removes it, its absence implies there was nothing to have a purpose in the first place, and so God does not exist. Heads, I win; tails, you lose.

Adam and Eve at Tanagra; Elohim, his arms wide

This objection is based on the notion that, according to Judaism and Christianity, God created the universe for humanity. This is incorrect. God created the universe for *himself*. In Judaism and Christianity God did not create the universe for us, meaning we are the most important thing. *God* is the most important thing, and he exceeds us infinitely. Most of his being is not about us at all.[19]

The reason for this misunderstanding, I think, is the biblical account of God's creation of human beings at the tail end of Gen 1. First, God creates human beings "in his image," which I'll deal with in chapter 10. Then he tells them that he has given them authority over all the other animals. This implies human beings are at the top of the ladder, and it's not much of a stretch to suggest that God created the earth, maybe even the entire universe, for them. And if God created the universe for them, for us, we can treat other forms of life, including animals, as mere objects with no intrinsic value and do with them whatever we want. This is called the *domination thesis*.[20] However, it was more a product of ancient Greek and Roman philosophy than Christianity. Aristotle claimed that "plants exist for the sake of animals and the other animals for the good of man . . . nature has made all the animals for the sake of men."[21] And, certainly, some Christian theologians followed the Greek and Roman views, at least to some extent, such as Aquinas, who based his theology on Aristotle. But Aquinas also acknowledged that the Bible forbade Christians from being cruel to animals, albeit not for their sake but to avoid developing a cruel disposition.[22]

However, the domination thesis is not the majority view in Christian history. That title goes to the *dominion thesis*.[23] The idea here is that the universe is God's, created for his purposes, and he created human beings in his image partially to act as caretakers of his creation. This does not preclude using animals for food or other purposes, but it does preclude treating them as mere objects. God created them and loves them, so we have a responsibility to them and him to treat them well.

19. C. S. Lewis, *Miracles*, 55.
20. Sampson, *Six Modern Myths*, 73–74.
21. Aristotle, *Politics* 1.8 (1256b.15–24).
22. Aquinas, *Summa Theologica* II-II, Q64, A1–2; *Summa Contra Gentiles* 3, C112.
23. Sampson, *Six Modern Myths*, 74–78.

In fact, the dominion thesis gives you a pretty good basis for environmentalism. I'm not even sure you can justify environmentalism without it or something very similar to it. And, irony of ironies, it was actually the *Enlightenment* and its pervasive mechanistic philosophy that reduced living creatures from beings with value to mere objects. But then when this became less popular among hoi polloi, the value/fact dichotomy was retroactively blamed on Christianity.[24]

Now, granted, plenty of individual Christians have treated animals horribly, plenty don't give a fig for the environment. Also granted, animal rights are mostly a contemporary phenomenon—I mean, if it took us until the nineteenth century to realize enslaving other people was evil, you've got to figure we wouldn't have recognized the evil of mistreating animals right away either. Those who did, however, were mostly Christians, such as Puritans, acting on Christian principles.[25] (The same goes for the abolition of slavery.)

As for the text in Gen 1, I think you can take it either way. You can reasonably take it as implying the domination thesis, and I don't think other biblical texts raised in this context rule it out either. But historically, Christianity has accepted the dominion thesis over the domination thesis. I don't think this is accidental; I suspect it's based on what the Bible says about God's nature and character and his love for the world he created.

Anthropic coincidence #2: The fine-structure constant

There is a second issue dealing with the Anthropic Principle that I want to mention. Scientists such as Carl Sagan and Stephen Hawking, and philosophers such as Nicholas Everitt, have argued that if God exists—a God roughly equivalent to the God of the Bible—then we should expect the universe to consist only of the earth, sun, moon, and maybe the planets.[26] In other words, we should expect something like the Ptolemaic universe. Anything more than that is gratuitous and unnecessary for

---

24. Sampson, *Six Modern Myths*, 71–73, 85–87. For an example of this, see Lynn White, "Historical Roots of Our Ecological Crisis."

25. Sampson, *Six Modern Myths*, 80–85.

26. Sagan, *Pale Blue Dot*, 26–39; Hawking, *Brief History of Time*, 133; Everitt, *Non-Existence of God*, 215–28. Thanks to Victor Reppert for helping me find the latter reference. For several similar comments by scientists, see Vainio, Cosmology in Theological Perspective, 107–9.

God's purposes. In a small universe, we can more easily believe that every element is there for humanity's sake.

Ignoring everything else I've already said about this, we can enhance this claim with a point made by Roger Penrose. He shows that the odds of our solar system forming randomly are insanely remote: 1 in $10^{10^{60}}$.[27] That's the number 1 followed by 10^{60} zeroes. This applies equally to the probability of a universe arising from chance that consists of just our solar system. This bolsters up Sagan and Hawking: if it's that improbable that a solar system universe would be produced by random processes, it is much more likely that, *if* it were produced, it would have been *intentionally* produced by an agent that wanted to create a universe like that. But of course, we don't live in that universe; we live in a universe orders of magnitude bigger than Ptolemy thought and so a) a solar system like ours was bound to form and b) the vast, *vast* majority of it is irrelevant to our existence—or, perhaps more significantly, we are irrelevant to the vast, vast majority of *it*.

So the improbability that random processes would produce our solar system and by proxy Ptolemy's universe—one chance in $10^{10^{60}}$—is so extreme that it would convince people like Sagan and Hawking that God exists, since it would indicate the universe is not here by chance but exists for our sake. A vast universe like our own, to which we are so irrelevant, is more plausibly the product of natural processes.

However, Penrose also shows that, for life to exist, the fine-structure constant, which measures the electromagnetism between particles, must be precisely what it is to within one part in $10^{10^{123}}$.[28] Now obviously 123 is 63 more than 60. But one part in $10^{10^{123}}$ is not 63 times more precise than one part in $10^{10^{60}}$; nor is it 10^{63} times more precise; it is $10^{10^{63}}$ times more precise. $10^{10^{60}}$ is to $10^{10^{123}}$ as 1 is to $10^{10^{63}}$. That's clear, right?

To put this into some perspective, there are approximately 10^{80} atoms in the entire universe. If we had 10^{43} universes the same size as our own (about 10 million trillion trillion trillion universes), they would collectively have 10^{123} atoms. If you converted all those atoms in all those universes into zeroes, that is how many zeroes would have to follow the number 1 to write out $10^{10^{123}}$. If just one of those 10^{43} universes had one fewer atom, then they could not represent how precise the fine-structure constant must be for life to exist. In contrast, to write out $10^{10^{60}}$ would

---

27. Penrose, *Road to Reality*, 726–32, 763–64.
28. Penrose, *Road to Reality*, 726–32, 762–65; Craig, *Reasonable Faith*, 148.

require us only (!) to convert one atom out of every one billion trillion atoms in our universe alone to zeroes (about 0.0000000000000000001 percent of all the atoms in the universe). Penrose uses a technical scientific term to contrast these two numbers: $10^{10^{60}}$ is "utter chicken feed" compared to $10^{10^{123}}$.[29]

In *Contact*, his only book of fiction, Sagan has two characters discuss what it would mean to discover a numerical pattern in pi. One asks, "If you look at enough random numbers, won't you get any pattern you want simply by chance?" and the answer is "Sure. But you can calculate how likely that is. If you get a very complex message very early on, you know it can't be by chance." The character goes on to say, "This is the only thing that would convince a skeptic."[30] Well, a condition present at the universe's inception that must be precise to one part in $10^{10^{123}}$ strikes me as "a very complex message very early on."

This is one of those places where it sounds like I'm trying to defend theism, but my point here is not an apologetic one. You could take both numbers as pointing to God or neither as doing so (perhaps you think probability cannot make the point). Or, theoretically, you could say one part in $10^{10^{60}}$ is not fine tuned enough to point to God, but one part in $10^{10^{123}}$ is—the cutoff point being somewhere between those two numbers. But the one thing you *can't* say is that one part in $10^{10^{60}}$ is so fine tuned that it points to God, but one part in $10^{10^{123}}$ is not, despite being insanely more extreme.

But this last option is essentially what Sagan and Hawking avail themselves of.[31] They thought a universe like Ptolemy's, which has only one chance in $10^{10^{60}}$ of being produced by random processes, is what we should expect if theism is true. But we don't live in that universe. We live in a universe where the fine-structure constant is exactly what it must be for life to exist, a single condition $10^{10^{63}}$ times more precise than Ptolemy's universe would have to be. So the actual universe is *much* more fine tuned than a universe that would convince Sagan and Hawking that God exists.

---

29. Penrose, *Road to Reality*, 764.
30. Sagan, *Contact*, 416, 419.
31. Along with anyone else who says that the modern discovery of the immensity of the universe points away from God and religion.

IV. SIZE DOESN'T MATTER

In the film *Contact* (not the book), there is a line that is repeated several times. One character asks another if there are any extraterrestrials out there in the immense universe, and the response is something to the effect of, "Well, if it is just us, it seems like an awful waste of space." Why? Well, because of the points we just went over: the vast majority of the universe is completely irrelevant to our existence, and our existence is completely irrelevant to the vast majority of the universe. It only makes sense that there are other forms of life out there.[32]

Whatever else is wrong with this claim, we've already seen that this isn't an issue since God delights in whatever he creates. To say that all that universe is a waste of space ignores this point. Moreover, waste is only an issue when there's scarcity. An unlimited, omnipotent God would not need to keep an eye on his expenditures when creating the universe unless he had some other motive for doing so.[33]

Of course, our irrelevance to the universe certainly prompts feelings of inadequacy in the face of ultimate reality. This is not unique to us: in the first century BC, Cicero used the vastness of the cosmos and the earth's diminutiveness to argue that the earth and humanity are insignificant,[34] and others followed suit. But what exactly is the argument? That bigger things are inherently more important than smaller things? I think that we tend to feel this is the case—again, in many languages, the word for "great" (as in *important*) starts off as the word for "big"—but we can't work from this to an actual argument about value or significance.

> There is no doubt that we all *feel* the incongruity of supposing, say, that the planet Earth might be more important than the Great Nebula in Andromeda. On the other hand, we are all equally certain that only a lunatic would think a man six-feet high necessarily more important than a man five-feet high, or a horse necessarily more important than a man, or a man's legs than his brain. In other words this supposed ratio of size to importance feels plausible only when one of the sizes involved is very great. And that betrays the true basis of this type of thought. When a relation is perceived by Reason, it is perceived to hold good universally. If our Reason told us that

---

32. See ch. 11.

33. Morris, *Logic of God Incarnate*, 77–78; Reppert, *C. S. Lewis's Dangerous Idea*, 124.

34. Cicero, *Dream of Scipio* 8.

size was proportional to importance, the small differences in size would be accompanied by small differences in importance just as surely as great differences in size were accompanied by great differences in importance. Your six-foot man would have to be slightly more valuable than the man of five feet, and your leg slightly more important than your brain—which everyone knows to be nonsense. The conclusion is inevitable: the importance we attach to great differences of size is an affair not of reason but of emotion—of that peculiar emotion which superiorities in size begin to produce in us only after a certain point of absolute size has been reached. We are inveterate poets. When a quantity is very great we cease to regard it as a mere quantity. Our imaginations awake. Instead of mere quantity, we now have a quality—the Sublime.[35]

Our feelings of insignificance in light of the size of the cosmos are based on our commonsense standard of measurement. If we really wanted to make it into some kind of argument, especially if it's supposed to be a *scientific* argument, we would have to use an objective scale of measurement, and the only one available is the logarithmic scale, basing it on the smallest and largest things in the universe. I'll use base 10 and meters but you could use any base number and any unit of measurement. The smallest thing is the Planck length at 10^{-35} meters. The largest thing is the universe itself at about 10^{25} meters. The earth's diameter is about 10^7 meters, while the human being is approximately 10^0 meters.[36] This puts the earth about 70 percent of the way up the scale and human beings about 60 percent of the way up. So using an objective standard of measurement doesn't communicate the same sense of our spatial insignificance.[37]

Of course, the response should be another "So what?" This doesn't assuage our sense of how insignificant we are, nor should it. But it's not trying to. It's just showing that we can't convert the feelings we have of our own irrelevance and insignificance into anything like a scientific argument. And even if we could, it would be making a point that no one has ever denied: that compared to the vastness of the universe, we are insignificant.

---

35. C. S. Lewis, *Miracles*, 56–57; emphasis in original.

36. Any number to the zeroth power is 1. Most of us are taller than one meter, but we're closer to that than we are to ten meters (10^1).

37. James Hannam, "Are We Very Big or Very Small?," https://bedejournal.blogspot.com/2006/12/are-we-very-big-or-very-small.html.

V. STRAW MEN

To review, Sagan and Hawking have both said or implied that, if God exists, we should expect the universe to have the dimensions of the Ptolemaic universe or a close proximity thereof. Why? Well, because there wouldn't be as much empty space going to waste. Not to mention that our planet meeting the necessary conditions for life would be much more improbable in a small universe like that rather than ours, and so it would best be explained as the action of a supernatural deity who made the universe just for us.

As noted, this is wrong on multiple levels:

- It assumes, falsely, that the Ptolemaic cosmology is the biblical cosmology.
- It assumes, falsely, that Ptolemy's universe is objectively small.
- It assumes, falsely, that the size of Ptolemy's universe makes the earth the biggest and most important thing in it.
- It assumes, falsely, that relative sizes are how God determines something's value.
- It assumes, falsely, that God created the universe for us rather than for himself.

So Sagan and Hawking are criticizing Christianity for something Christianity has never said. They are judging its plausibility by constructing a straw man, and then contrasting this straw man with the claims of science.

John Shelby Spong was an Episcopalian minister who argued vehemently against the traditional understanding of Christianity, God, Jesus, the Bible, etc. One comment he repeated is about the scientific absurdity of Jesus' ascension: after he was resurrected, Jesus rose into the air out of sight, presumably going to heaven. Spong tries to argue that this might have made sense when people thought that heaven was just on the other side of the clouds, but now that we have a clearer idea of the size of the universe, it's scientifically absurd. We've already seen that they *didn't* believe that, but just ignore that for the moment. In making this point Spong refers to a conversation he had once with Sagan, who told him that if Jesus had ascended away from the earth at the speed of light, he would still be in the Milky Way Galaxy today.

Yet the *South England Legendary* from the thirteenth century expresses a similar sentiment. It says that if a man could travel forty miles a day, about as fast a speed that people could travel at the time, it would take him eight thousand years to reach the sphere of the fixed stars.[38] If Jesus had ascended at that speed, he wouldn't even be close today, two thousand years later. Even if he'd ascended at three times that speed, he'd still be well within the Ptolemaic universe. Yet, somehow, medieval Christians didn't think Jesus' ascension was still going on twelve hundred years later—or even twelve seconds later. It was accomplished. Whatever else you might think about the ascension, however implausible you may think it is, the size of the universe doesn't add anything to the equation. They already knew perfectly well that the universe was so large that for Jesus to physically, linearly travel to its periphery would have taken a ridiculous amount of time. Since it didn't take *any* amount of time, they knew that the ascension didn't involve physical, linear travel.

This shouldn't be surprising. It would have been a miracle. No one suggested it would have been a physical, natural event, it was a supernatural event. *Of course* it wouldn't have involved physical travel, and the ancient and medieval Christians always knew it didn't. Someone tell Spong.

---

38. D'Evelyn and Mill, *Text*, 418; C. S. Lewis, *Discarded Image*, 98; cf. Maimonides, *Guide for the Perplexed* 3.14.

Chapter 5

Medical History

I. EARLY MEDICINE

I'VE IMPLIED A COUPLE of times that proto-science, premodern science, wasn't as horrifically crazy as we tend to make out. Astrology and alchemy, for example, made sense on some level, they just turned out to be incorrect. When it comes to medicine, that goes out the window. Some aspects of the history of medicine are freaking *terrifying*. Gary Ferngren, a historian of medicine, writes, "It is doubtful whether a sick person gained much biomedical advantage from seeing a physician until less than two centuries ago."[1] Hannam claims that, until fairly recently, if you got sick, your best bet was asking a priest or pastor to pray for you, and second best was going to a local healer who mostly prescribed folk-medicine ointments that would do neither good nor harm. But if you went to an actual doctor, they would prescribe more invasive procedures that tended to make you less healthy.[2] I should probably qualify this by saying I find some aspects of contemporary medicine freaking terrifying too.

Prior to the development of germ theory in the nineteenth century, the traditional take on human medicine was humoral theory. The idea here is that there are four bodily fluids: blood, yellow bile, black bile, and phlegm. Yes, I know it sounds gross. Your personality, or temperament, is based on the balance of these fluids and too much in one direction meant

---

1. Ferngren, *Medicine and Religion*, 11.
2. Hannam, *God's Philosophers*, 109–11.

they were out of sync. Being too manic meant you had too much blood in your body; too choleric meant too much yellow bile; too melancholic meant too much black bile; and being too phlegmatic meant too much (go on, take a guess) phlegm. The idea goes back to the pre-Socratics and was notably developed by Hippocrates and Galen. They also tied the four humors to the four seasons and the four elements.[3]

In this chapter, however, I will look at some urban legends about how Christianity supposedly stood in the way of the advancement of medical science. You may or may not have heard these myths before, but I often see them used to promote the "eternal struggle between science and religion" motif.

In her defense of eliminative materialism—which, roughly, is the claim that there are no such things as beliefs, there never have been and there never will be—Patricia Churchland recognizes that it may entail a complete overhaul of our "moral and legal institutions." To defend this possibility, she points to the science vs. religion meme: "It is at least conceivable that our moral and legal institutions will be seen by future generations to be as backward, superstitious, and primitive as we now see the Christian church's doctrine of past centuries concerning the moral significance of disease and the moral property of anesthesia, immunization, and contraception."[4]

Artificial contraception is an issue within Catholicism; Orthodox and Protestant Christianity don't have a problem with it as such. But what exactly does Churchland mean by "the Christian church's doctrine . . . concerning the moral significance of disease"? Or "the moral property" of anesthesia and immunization? How are disease, anesthesia, and immunization connected to Christian doctrine, and how are they moral issues at all?

Interestingly, Churchland doesn't even mention the most common claim of religion supposedly getting in the way of medical science. I don't know if this is because she disputed its authenticity, hadn't heard about it, or it just didn't fit into the flow of the text. As you should be able to tell by now, I don't care about the flow of the text, so here we go.

---

3. Ferngren, *Medicine and Religion*, 44–45; Jouanna, "Legacy of the Hippocratic Treatise."
4. Churchland, *Neurophilosophy*, 399.

II. THE BANNING OF HUMAN DISSECTION

When I was in high school, a teacher told my class that, prior to the Scientific Revolution, everyone thought the heart was in the abdomen instead of the chest. This was because you can feel the heart's pulse in your abdomen, and they never dissected human bodies since it was forbidden on religious grounds. It's only in the early modern era that people started *looking* into dead bodies and found out where everything is. I believed her. I never asked myself some obvious questions like "Can't we feel our pulse in multiple places in our body? Why didn't they think the heart was in the wrist or the neck, why privilege the abdomen?" or "Didn't they recognize the morphological similarities between human beings and other animals that they *did* dissect and infer that our organs were located roughly on par with theirs?" or "Wasn't warfare much closer up until fairly recently? Wouldn't people have seen their comrades and enemies sliced open and observe their internal organs? Wouldn't they have used this information to know how to cause the most damage and where to stick the spear?"

The story this teacher told us is that the Christian church was opposed to human dissection, even if it was for scientific reasons. Perhaps this was because the claim that we are created in God's image applies to the specific form of the human body, so that even when it has been de-souled it should not be treated as a mere object or means to some other end rather than an end in itself. Whatever the motivation, there are plenty of reasons to think that the church was strongly opposed to the dissection of human corpses. The conciliar decree *Ecclesia abhorret a sanguine* (The Church abhors the shedding of blood), written around 1250, explicitly forbade cutting up dead bodies. Half a century later, Pope Boniface VIII issued a papal bull, *De Sepulturis*, to the same effect. And medical scientists were persecuted and even executed, *martyred*, for their attempts to learn more about the human body via dissection in order to assuage human suffering, such as Andreas Vesalius and Michael Servetus, both in the sixteenth century. So, *yeah*, Christianity came out pretty strongly against human dissection.

Except, *nah*, none of that is true. The "image of God" was never thought to apply to a dead body, so there would have been no concerns from that direction. There might have been some concerns that it disrespected the form God chose for human bodies, but this would have come from Islam since Sharia law forbade dissection, although Muslim jurists never formed a consensus against it. "Human dissection was viewed in Islam as a violation of the human body, which was to be respected as God's most noble creation."[5] This idea might have influenced some Christians in the later Middle Ages and Renaissance when Islam was having an intellectual influence on Christian culture, but since Islam decries any concept of an image of God, Judeo-Christian doctrine did not enter into it.

As for that conciliar decree "The Church abhors the shedding of blood" that is often cited to make the point—well, it turns out it doesn't really exist. Track the footnotes and you don't get back to an original source.[6] Saying that God or the church "abhors the shedding of blood" was a common phrase in medieval Christianity, but there is no conciliar decree with that name, nor is there any other that bans human dissection. The papal bull *De Sepulturis* does exist, however, and it was taken by some people at the time to ban human dissection, or at least as requiring a papal injunction to do it.[7] In reality, it was banning a very specific practice regarding the treatment of corpses. It had become popular, for some ungodly reason, to cut up a dead body and boil the flesh off its bones. Perhaps this was so that the bones could be transported across long distances to be buried in the family plot. It was common among crusaders. Boniface was prohibiting that particular practice, not human dissection.[8]

Interestingly, the misinterpretation that it *did* prohibit human dissection was more popular the further one was from the Vatican: in Italy and the other Mediterranean regions, it was taken as banning that one practice, and human dissection continued unabated,[9] although it wasn't very common. When the universities of Padua and Bologna requested ecclesiastical permission to perform dissections in their medical schools in the fifteenth century it was immediately granted—and why wouldn't it be if it was never prohibited in the first place?[10]

So there wasn't any injunction against human dissection. There were decrees forbidding clergy from performing surgery on living bodies, not dissection on dead ones, but that was because a) it distracted them from the spiritual concerns they were supposed to be focused on; and b) they

5. Ferngren, *Medicine and Religion*, 126–27.
6. Katherine Park, "Myth 5," in Numbers, *Galileo Goes to Jail*, 43–49 (esp. 46).
7. Park, "Myth 5," in Numbers, *Galileo Goes to Jail*, 47.
8. Thadani, "Myth of a Catholic Religious Objection to Autopsy."
9. Park, "Myth 5," in Numbers, *Galileo Goes to Jail*, 47.
10. King and Meehan, "History of the Autopsy," 521.

often made things even worse than the doctors tended to make it.[11] Leave the doctoring to the doctors; you monks, just be all monkish.

What about the persecutions and executions? Well, the text that claims Vesalius got into trouble for dissecting a dead body is recognized as fraudulent by historians, so it probably never happened.[12] And even if the text *was* genuine, it doesn't say Vesalius got in trouble for dissecting a dead body. It says he got in trouble for dissecting a *mostly* dead body. In other words, it says he accidentally killed a guy by mistaking him for a corpse and cutting him up. Granted, the poor guy quickly *became* a corpse, but that wouldn't be a point in Vesalius's favor. Michael Servetus is a different matter since he really was executed by the Calvinist church. But it had nothing to do with his medical investigations and dissections, it was about his theological claims. I'm not defending his execution: obviously they shouldn't have killed him, I'm just saying it was unrelated to his predilection for human dissection.[13]

In fact, according to Katharine Park, no one was ever prosecuted for dissecting a corpse within Christendom during the Middle Ages, Renaissance, and early modern era.[14] It was a social taboo for several reasons, no doubt including the same reason it's taboo for most of us: it's gross. I doubt many of you would feel comfortable cutting up a dead body or even watching someone else do it. Most people would at least find it unnerving because it is not easy for us to dissociate the *body* from the *person*. Dissecting the body would feel like one was treating the person as a mere object. Abuse of a corpse is a crime in all or nearly all countries *today*, and it should be. "How we treat our dead is part of what makes us different from those who did the slaughter."[15]

In the fourth and third centuries BC, some human dissection was permitted, briefly, in Alexandria, but it was unheard of anywhere else.[16] Galen (second century AD), the primary source of our knowledge of Western medicine in antiquity, never did it, although, sweetheart that he was, he did publicly vivisect pigs and young goats (that is, dissect them while they were still alive).[17] In the Middle Ages, pigs were sometimes

---

11. Amundsen, "Medieval Canon Law on Medical and Surgical Practice."
12. O'Malley, *Andreas Vesalius of Brussels*, 304–5.
13. Park, "Myth 5," in Numbers, *Galileo Goes to Jail*, 48.
14. Park, "Myth 5," in Numbers, *Galileo Goes to Jail*, 47.
15. Tim Minear, *Firefly*, episode 3, "Bushwhacked."
16. Park, "Myth 5," in Numbers, *Galileo Goes to Jail*, 45.
17. Hankinson, "Man and His Work," 11–12.

dissected "because of the perceived similarity of their organs to those of humans."[18] As a regular practice, human dissection started around 1300. This was a watershed moment. "The introduction of human dissection in western Europe is one of the most surprising events in the history of natural science. It was practically unheard of in any other culture due to strong taboos against cutting up bodies and not giving them the proper respect demanded by tradition."[19] So, contrary to what that high school teacher told me, Christian Europe was pretty much the only place that was willing to dissect human bodies. The group she blamed for *not* doing it was the only group that *did* do it. "If the Catholic church had really objected strongly to human dissections, they would not have rapidly become part of the syllabus in every major European medical school."[20]

Moreover, there were many accepted reasons for cutting into dead bodies. Embalming to prepare for burial and autopsies were common enough, but the internal organs of noteworthy people were also inspected for signs of sanctity (like that they were not decomposing but still in pristine condition) on the off chance that they might be considered for sainthood. Or they might divide the body up to be sent to various places to serve as holy relics. When pregnant women died, they would perform caesarean sections to remove and baptize unborn infants.[21] Note that most of the reasons *for* cutting into bodies were specifically religious reasons.

To be sure, when dissection for medical research became more common, there were strong social and legal limits (not *religious* limits) placed on which bodies were eligible, and this led to a scarcity of available corpses, a dearth of death. This was mostly for issues concerning personal and family honor. It was considered disrespectful to the person and the family to dissect a dead body since it rendered the body incapable of having a proper burial. There were no similar objections to autopsies since they didn't have these consequences. For this reason, most bodies that were available for dissection were those of criminals or foreigners—where "foreigners" refers to people from more than thirty miles away.[22]

There *was* a moral problem involved, it just wasn't about cutting into dead bodies. It was about the *acquiring* of dead bodies—specifically, grave robbing, which was forbidden by secular and ecclesiastical

---

18. Ferngren, *Medicine and Religion*, 101.
19. Hannam, *God's Philosophers*, 252.
20. Hannam, *God's Philosophers*, 255.
21. Park, "Myth 5," in Numbers, *Galileo Goes to Jail*, 44–45.
22. Park, "Myth 5," in Numbers, *Galileo Goes to Jail*, 48–49.

authorities, as it is today. It got so bad that some funerals were hijacked (I mention this in case you're tempted to think the present era is more immoral than others).[23] Another issue was that some of the people trying to acquire dead bodies *weren't scientists*. They were artists who wanted to procure dead bodies in order to make detailed drawings of the parts.[24] They never got in trouble for it either, although people may have been more reticent to sell a corpse to someone who said, "Oh, no, I'm not a scientist, I just want to cut it up and look at it."

III. BANNING ANESTHESIA DURING CHILDBIRTH

Above, I quoted Churchland and her reference to "the moral property of anesthesia." Here's the dealio: James Young Simpson discovered that chloroform could be used on human beings in 1846. He and some other doctor friends would regularly experiment with drugs to search for potential anesthetics (sure, that was it), and when chloroform night rolled around, they lucked out and took exactly the right amount to not die. But Simpson immediately saw that you could use chloroform to knock people out for especially painful procedures. Since listening to disco music wasn't an option yet, they had to settle for the rigors of childbirth.

However, we are told that Christians immediately objected to using it this way. Pain in childbirth is part of the curse of the fall of humanity, so alleviating that pain is an attempt to thwart the righteous punishment of God. The biblical passage in question is Gen 3:14: "I will make your pains in childbearing very severe; with painful labor you will give birth to children." Trying to lessen or even *erase* the consequences of sin is to take away one of the main motives to refrain from sin in the first place. Trying to erase the consequences of *original* sin is essentially to embrace the sin of trying to be like God. Simpson was a Christian and he denied any conflict between science and religion, affirming that the Bible was substantially correct where it touched on scientific and specifically medical issues. So in late 1847 he wrote a tract to answer the theological

---

23. Park, "Myth 5," in Numbers, *Galileo Goes to Jail*, 48.

24. Leonardo DaVinci was able to procure body parts, but he is often considered a scientist as well (and rightly so, in my humble yet correct opinion) because of what he wrote in his notebooks. But, according to Hannam, these notebooks were unknown in his time and he had "no impact on the development of western science at all" (*God's Philosophers*, 7). At any rate, his scientific endeavors had nothing to do with his fondness for corpse acquisition.

objections to using chloroform in childbirth.[25] Clearly, then, the Christian reaction to Simpson's discovery was widespread, and of course his tract only influenced the few who actually read it.

Except there were no widespread theological or religious objections to using chloroform for childbirth. Simpson wrote his tract in late 1847, and by mid-1848 he said he wasn't hearing objections anymore.[26] In fact, it's entirely possible that his tract was not intended to answer objections that had been made but rather to *forestall* potential objections.[27] Nevertheless, the claim of widespread objections on theological grounds is a myth.[28] Some people said they had theological *questions* about it before reading Simpson's tract but having questions and even concerns is not the same thing as having objections. This is not to say that no one objected to it, but they were few and far between, and were reported only secondhand.

> With only one exception, no reference at all has been found to the existence of any religious objections to anesthesia either in Europe or America. . . . Both America and Britain shared an extensive period of evangelical revival during the nineteenth century. The fact that the social climate of neither country gave rise to religious objections to anesthesia, other than by a very few individuals, suggests that such views were not widespread.[29]

The alleged objection, again, was that making childbirth more bearable (ha!) is to erase the righteous curse for sin. But women are not the only ones cursed by original sin. Men are too. "Cursed is the ground because of you; through painful toil you will eat food from it all the days of your life. It will produce thorns and thistles for you, and you will eat the plants of the field. By the sweat of your brow you will eat your food until you return to the ground" (Gen 3:17–19). So if alleviating the pains of childbirth goes against Eve's curse, wouldn't labor-saving devices go against Adam's curse? Wouldn't just clearing land go against the curse that thorns and thistles will hamper our attempts at agriculture? Yet this practice has been around for a while without anyone noticing any horrifying theological implications.[30]

---

25. Rennie B. Schoepflin, "Myth 14," in Numbers, *Galileo Goes to Jail*, 123–30 (esp. 124).
26. Schoepflin, "Myth 14," in Numbers, *Galileo Goes to Jail*, 124–25.
27. Farr, "Early Opposition to Obstetric Anesthesia," 905–6.
28. Farr, "Religious Opposition to Obstetric Anesthesia."
29. Farr, "Religious Opposition to Obstetric Anesthesia," 176.
30. Schoepflin, "Myth 14," in Numbers, *Galileo Goes to Jail*, 127.

There was some opposition to the use of chloroform, but it came from medical professionals, and was on medical and scientific grounds. Some thought it went against natural law, not the Bible or Christianity.[31] Roughly, natural law as applied to medicine is the idea that things are the way they are for a reason, and to ignore this indiscriminately is unwise. The reason why we experience pain is to warn us that the body is being damaged. To prevent that pain could very easily become dangerous. There are people who have Congenital Insensitivity to Pain (CIP). Most of them don't live to adulthood. Feeling pain is critical for survival. In an interview, the late, great science-fiction author Gene Wolfe said that pain not only didn't get in the way of his becoming a Christian as an adult, but that he considered it a positive reason to believe in God.

> **You once said that pain tends to prove God's reality rather than the opposite; that pain was not a theological difficulty for you.**
>
> No, it isn't. If you catch a dragonfly and bend the end of its body up, it will eat itself until it dies. When people have had their mouths numbed for dentistry, they must be warned not to chew their tongues. I think if we assume that pain is simply an evil we're oversimplifying things.
>
> **[Thinks a moment.] You're saying that pain may be a necessary design feature that the Divine Engineer—**
>
> Yes, absolutely.
>
> **—put into his animated machines.**
>
> If you had living things without pain, they would have a very rough time surviving.[32]

So the objection is that, if it's natural to experience pain, then to meddle with it may end up causing more harm than good—and since the charge to doctors is "first do no harm," this could be problematic.

IV. BANNING VACCINATIONS

Churchland also referred to "the Christian church's doctrine of past centuries concerning the moral significance of disease and the moral

---

31. Farr, "Early Opposition to Obstetric Anesthesia," 896–902; Schoepflin, "Myth 14," in Numbers, *Galileo Goes to Jail*, 127–28.

32. Pontin, "Q&A with Gene Wolfe."

property of... immunization." This is a reference to the claim that Pope Leo XII banned smallpox vaccinations in 1829. The supposed objection here is similar to the supposed objection to chloroform, but without the supporting Bible passage: disease is a righteous and appropriate punishment from God. As such, vaccinating against diseases is an attempt to thwart God's will. Specifically, Leo said, "Whoever allows himself to be vaccinated ceases to be a child of God. Smallpox is a judgement from God; thus vaccination is an affront to Heaven."[33] This is similar to Father Paneloux in Albert Camus's *The Plague*, who responds to the titular pestilence by saying it's a righteous judgment from God. When an innocent child dies from the disease, Paneloux says that God must have willed it, in which case we should will it too. Fortunately for all the other characters, as well as the reader, Father Paneloux later dies of the plague himself.[34]

Anyway, the background here is that one person in three who caught smallpox died of it. There had been a practice of inoculating people with a small amount of the smallpox virus, advocated by Christian leaders such as Jonathan Edwards, but this obviously carried huge risks. People respond differently to these things, so it was very precarious. While the survival rate was significantly higher than catching smallpox, the patient still ran the serious chance of dying from it (Edwards did, in fact). So inoculating people with smallpox was a risky solution.[35]

However, there was a similar disease that caused some of the same symptoms on a much lesser scale and was not particularly dangerous: cowpox. This was primarily a disease of animals—like, you know, cows—but very fortunately and providentially human beings can catch it. The people who *did* catch it were people like milkmaids who worked with certain types of animals—like, you know, cows. Edward Jenner was a physician who noticed that people who caught cowpox didn't catch smallpox, so he started vaccinating people with cowpox.[36] The word "vaccine" and its derivatives come from the Latin word for "cow."

So what about good old Pope Leo XII banning smallpox vaccinations in 1829? Well, of course he didn't. The quote of him condemning

---

33. McCormick, *Health and Medicine in Catholic Tradition*, 17.

34. As with much of his work, Camus's point is that life is absurd and has no meaning, so we should make our own meaning. We should believe that the world is objectively meaningless, but subjectively meaningful. It's not easy to make sense of this. I know a philosopher who became a Christian after reading *The Plague*.

35. Boylston, "Origins of Inoculation."

36. Riedel, "Edward Jenner and the History of Smallpox and Vaccination."

it above is a fabrication. I got it from Richard McCormick, who gets it from Louis Janssens, who gets (and translates) it from Abel Jeannière, who does not provide a source for the quote. Donald Keefe contacted Jeannière to ask him about it, and he pointed him to another article of his which further points to a book by Pierre Simon, written in 1966, which provides no reference itself. Ultimately, Jeannière defended it by saying Leo had called vaccinations "bestial" before becoming pope.[37] The earliest reference I've seen accusing Leo of banning vaccinations is the introduction to G. S. Godkin's *Life of Victor Emmanuel II*, published in 1880.[38]

> He was a ferocious fanatic, whose object was to destroy all the improvements of modern times, and force society back to the government, customs and ideas of mediæval days. In his insensate rage against progress he stopped vaccination; consequently, small-pox devastated the Roman provinces during his reign, along with many other curses which his brutal ignorance and misgovernant brought upon the inhabitants of those beautiful and fertile regions.

Basically, this is just made up.[39]

There *was* some backlash that sometimes took the form of religious objections to vaccinations against smallpox, but it tended to be personal decisions, not widespread protests. It was primarily a working-class issue about bodily autonomy: the argument was that people should not be compelled by force of law, and by the cultural elites, to do something to their bodies they don't want to do. Yes, people sometimes invoked religious concerns (even including "God ordains disease"), but the objections were convoluted and secondary to the primary concern.[40]

To be sure, Christianity has a long tradition of *praying* for the alleviation of disease, as do many other religions, and this comes with the belief that God can and sometimes does supernaturally heal people. But this does not conflict with the use of standard medical procedures, nor has it historically been taken as such. Hospitals are a Christian invention,

---

37. McCormick, *Health and Medicine in Catholic Tradition*, 17; Janssens, "Artificial Insemination," 11; Jeannière, "Corps malleable," 94; Simon, *Contrôle des naissances*, 164; Keefe, "Tracking a Footnote."

38. Godkin, *Life of Victor Emmanuel II*, xiv. I found this reference via Humphrey Clarke, "Pope Leo XII and the Vaccination Ban," https://bedejournal.blogspot.com/2009/03/pope-leo-xii-and-vaccination-ban.html.

39. Bercé and Otteni, "Pratique de la vaccination antivariolique."

40. Durbach, "They Might as Well Brand Us."

after all.[41] Naturally, there will always be people who take an attitude that if God wanted us to take medicine, he would have made it directly—which makes as much sense as saying if God wanted us to farm he would have made tractors—but this is simply on the level of personal attitude, not denominational doctrine.

This issue is relevant today because we have a lot of people trying to opt out of vaccinating their kids, and the opting-out option often takes the form of requesting a religious exemption. But the religions in question are just local churches, synagogues, or religious communities, not denominations (plus, I'm pretty sure most people have nonreligious motives for wanting to not vaccinate their children and are using the religious exemption as an excuse). With the recent travails of COVID there were some people who objected to being vaccinated for reasons pretty close to those of the working class in mid-nineteenth-century England. At any rate, the point is that Christians didn't have a problem with vaccinations historically and those that do today are not basing it on common theological positions or interpretations of the Bible. In the same way, some rural Muslims, especially in North Africa, reject vaccinations, but not for anything in the Qur'an or official Islamic teaching—they think vaccines are part of a Western plot to harm them.

Of course, there *are* some religious groups who are opposed to vaccinations, surgeries, and various medical treatments. Christian Scientists (who are neither) think the physical world is an illusion and thus so is physical disease. Scientology rejects mainstream psychiatry, including medical treatment for psychiatric malfunctions. But such ideas are not representative of religion in general regarding medicine, despite widespread claims to the contrary.

V. ONE MORE ISSUE

So where did Churchland get the idea that there was not just resistance to chloroform and vaccinations within Christian circles but that they contradicted Christian *doctrine*? Fortunately, she tells us (such is the glory of parenthetical references). She got it from Andrew Dickson White's 1895 book *History of the Warfare of Science with Theology in Christendom*, a book that has been recognized as a work of propaganda for decades.[42]

---

41. Ferngren, *Medicine and Religion*, 113–15.
42. See below, 191.

This shows once again how people are selectively skeptical. Madalyn Murray O'Hair and her followers uncritically accepted a silly story that no one knew sex led to pregnancy until the nineteenth century.[43] Why? Because it bolstered up their preconceived ideas about religion. Churchland uncritically references a text that has long been rejected by historians as a reliable source. Why? Because it bolsters up her preconceived ideas about religion.

However, there was one point she made that has some teeth to it: contraception. Artificial birth control is a recent invention, and the Catholic Church has come out in opposition to it. Protestantism and Orthodoxy, however, allow it, and the Catholic Church still allows methods based on women's fertile cycles, like the Creighton and the symptothermal methods.[44] Unfortunately, these have been lumped together with the older and spectacularly ineffective rhythm method.

In response, it needs to be pointed out that *this issue has nothing to do with whether science and religion conflict*. It doesn't deny or challenge any scientific claim. It is an ethical issue about what we *do* with certain scientific advancements. Regardless, the Catholic motive for rejecting artificial birth control is not an unthinking reaction or a power play to control people. It is a very well-thought-out position based on natural law. If you doubt this, read the 1968 encyclical *Humanae Vitae*.[45] This doesn't mean that you have to agree with it. I don't. For that matter, many Catholic cardinals and bishops (and laity) strongly disagreed with it when it came out. But if you think it's a knee-jerk reaction or that they have nefarious ulterior motives, you first have to see what *they* say is their motive. Otherwise, *you* are the one with the knee-jerk reaction or basing your take on this issue on ulterior motives. Not to mention the minor point that you'd be committing an ad hominem fallacy—criticizing the person (or in this case, institution) instead of the argument, something that has been recognized as sloppy thinking for as long as people have been thinking. Of course, people believe things for the wrong reasons, even *immoral* reasons, all the time, but if what they say is true, their motives are simply irrelevant. Two plus two doesn't stop equaling four just because your high school bully believes in math.

---

43. See above, 13–14.
44. Thanks to Patrick Tomassi for pointing this out to me.
45. Paul VI, "*Humanae Vitae*."

Chapter 6

Tales of the Righteous Dead

STORIES ARE IMPORTANT. THE story of Columbus standing alone against the religious hierarchy is a story intended to inspire people to think for themselves, and does so by portraying religion as the bad guy to Columbus's good guy persona (although calling Columbus the good guy has been challenged of late). We've already seen that this story is false, but it follows a pattern: the lone scientist persecuted by the evil church, sometimes to the point of death. Martyred in the name of science, knowledge, reason, and truth.

Martyrs make great stories. They are those who prefer death to compromising their principles. They are admirable people, people we should emulate, and their deaths give an emotional punch to the positions they died for. Once someone dies for a cause, that cause often becomes much more popular: If the hero thought the cause was more important than their own life, who am I to behave otherwise? Unfortunately, this also creates a motive for altering details to make the story fit the martyr motif when the actual account wouldn't make the case as powerfully—or at all. Altered details often include who the bad guy is, what the bad guy's *motivation* is, or what the point of contention was between the good guy and the bad guy. Not to mention whether the good guy and the bad guy can be painted so one-dimensionally by calling them "the good guy" and "the bad guy."

I. SOCRATES

The most famous martyr along these lines, no doubt, is Socrates. He was pretty impressive: he served in the Athenian military into his fifties, fought in numerous wars, and was known for being calm and fearless in the heat of battle. Bear in mind this was a time when dying in battle was a prolonged event: you'd be run through with a sword or spear, fall in the mud, and then die of thirst after a few days of unbearable agony. Supposedly, the Spartans would sometimes retreat rather than fight him face-to-face. He was also the first person over the age of six to continue asking, "Why?" and "How do you know?" to everything everyone said. Apparently, he was also butt-screaming ugly: his nickname was Frogface.[1]

A childhood friend of his visited the oracle at Delphi (basically, a conduit for the gods) and asked if Socrates was the smartest man. The oracle responded that no one living was smarter than Socrates. When word got back to him, Socrates was utterly bewildered by this: the only thing he thought he could say he knew was that he knew nothing. Eventually he realized that was *why* no one was smarter than him. Everyone else thought they knew a lot of things, but they were wrong; only Socrates knew that he didn't know anything. Thereafter he believed he was on a mission from God to spread the gospel of ignorance.[2]

This isn't nearly as snarky as it sounds. He thought that being in connection with reality was intrinsically good, and being out of connection with reality was intrinsically bad. Believing something that is false is to be out of touch with reality, since you think reality is a way that it is not. So showing people that they didn't know squat was one of the greatest goods you could do for them, even if it was a painful or excruciating experience. To this end, he would ask them questions that forced them to dig deeper and deeper until they realized that they didn't know what they claimed to know. Sometimes they thanked him, sometimes they insulted him, sometimes they cried.

If you doubt that Socrates was right about this, let me give you an example I read somewhere. Say there's a guy, let's call him Bartholomew Squishington, who's a complete moron and so gets into an Ivy League school. Bartholomew writes a paper for his physics class that is so bad his professor can't help but laugh at it, like *really* laugh. But then he has an idea: he gives the paper an A+ and tells Bartholomew it's the most

---

1. Herrick, *Think with Socrates*, 7, 10–11.
2. Plato, *Apology* 20e–23b.

brilliant thing he's ever read. He suggests he send it to a physics journal. Then he calls the journal and lets them in on the joke, so they go along and publish the paper; all the while, everyone is laughing at this poor guy. On the strength of this, Bartholomew Squishington becomes a physics major, graduates summa cum laude, goes into graduate work, earns his PhD, and gets a primo academic position, writing and publishing everything that comes into his head. All of it is crap, though; the entire physics community is in on the joke and laughing their heads off at Bartholomew behind his back; everything he writes is just irredeemable nonsense. Eventually, he dies peacefully, thinking he has made the world a better place and was a valued and esteemed member of society. But again, he was not; he was a laughingstock and the whole thing was a practical joke at his expense.

So the question is, would you want to live this life? Bear in mind, *you'd never find out it was bogus*, you'd die thinking it was all real. If you'd be OK with it, ask yourself if you'd want a life like this for your children. And even if you'd be OK with that, ask whether you would prefer that life or an identical life except where everything was *real*, where you really *were* a brilliant physicist, and really *did* make the world a better place. If you'd prefer the life where the accomplishments and accolades were genuine over the life where it was all a practical joke, you recognize that it is better to be connected to reality than not, which is Socrates's point.

Socrates's penchant for asking inconvenient questions (the aforementioned "Why?" and "How do you know?") got him in trouble with those in favor of maintaining the status quo. They accused him of discouraging worship of the traditional Athenian gods and of corrupting the minds of the young.[3] He denied both charges. For one thing, encouraging people to engage critically with their beliefs is pretty much the opposite of corrupting them, and for another, asking questions isn't to deny anything, and he didn't deny the gods of Athens or encourage anyone to withhold worshiping them. He did suggest, however, that there is a meta-God over all of them, namely, the God who told him to go into the world and preach the good news of how we all don't know anything. He owed Athens a great deal, he said, but he owed God more, and God told him to teach.[4]

---

3. Plato, *Euthyphro* 2c–d.
4. Plato, *Apology* 37e.

He was given a trial before the Senate, found guilty, and sentenced to death. This wasn't *too* disconcerting because the system was such that the Senate would sentence you to something extreme and you would suggest an alternative form of punishment. The risk was if your suggestion wasn't severe enough, the Senate's suggestion would be taken instead, so you had to propose a punishment that was unpleasant enough that they'd be willing to accept it. So they sentenced Socrates to death, and it was up to him to suggest a different sentence, one that the Senate would be willing to agree to instead.

He suggested that he be provided with free meals in the public dining hall for the rest of his life to thank him for his service to Athens by teaching the public.[5] This was the standard reward for a benefactor of Athens. Oddly, the Senate decided to stick with the death penalty.

He spent the last month of his life in prison being visited by his friends and philosophizing with them. On the day of his execution, he debated with some of his students about whether there is life after death.[6] He thought there was. But, holy *crap*, that's the sign of a true philosopher if anything is: Being open to having a forthright discussion like that right before you die?[7] And while we could criticize his students for the insanely inopportune act of trying to convince a man about to be killed that there's no afterlife, it fits in well with Socrates's view that it is intrinsically better to know a horrifying truth than to believe a comforting falsehood.[8] He was forced to drink hemlock and died ("forced" in that he wouldn't have drunk it in other circumstances; not that they held him down and poured it down his throat).

Socrates could have gone into exile. He refused because to do so would be to disobey the God who told him to teach. He could have escaped from prison—even his jailer was willing to help him do so because

---

5. Plato, *Apology* 36d.

6. Plato, *Phaedo* 70a–103e.

7. Another example is İlham Dilman who, upon discovering he was not long for this world, decided to spend his final months reading and critiquing books by philosophers he had never read but knew he would disagree with. The result is his *Philosophy as Criticism*.

8. There's a similar passage at the beginning of Nietzsche's *Thus Spoke Zarathustra*, where Zarathustra tells a dying acrobat there is no afterlife, and his whole life had been a lie. The acrobat thanks him profusely for putting him in touch with reality, and then dies.

he recognized Socrates was a righteous man. But Socrates refused because he owed too much to Athens to avoid her requirements upon him.[9]

None of the above account is incorrect, and I think it's pretty clear from it that Socrates died for a cause. He was a martyr. The question is, what was he martyred *for*. Obviously it had nothing to do with Christianity since Socrates was killed four centuries before Jesus. But one of the reasons his opponents gave for his execution was that he discouraged worship of the traditional Athenian gods. He denied this, and in fact his last words were to ask one of his disciples to make an offering to one of the gods on his behalf.[10] But the charge against him indicates that one of the motives for executing him was a religious one. So while Socrates might not have been martyred by Christianity specifically, he was martyred by religion.

This shouldn't be surprising: people who go against the reigning ideology—be it religious, political, social, or even economic—can be persecuted and killed for it by the ideology's supporters. So, of course, religion has been used in this way, and plenty of religions even encourage it. I'm not denying this at all, I would only point out that a) it's not something particular to religion, since it applies to ideologies in general, and b) the fact that *some* religions encourage it does not mean that *all* religions do—and it's very relevant to ask whether people who persecute and kill in the name of their religion are being consistent with the content of their religion. In other words, the fact that people do such things doesn't necessarily mean that the religion or ideology in question is at fault. People can be hypocrites.

Having said this, I think a good case could be made that the real charge against Socrates was that he was challenging the status quo and making people question the cultural authorities—and he was doing this at an especially dangerous time. To stop him, they charged him with whatever could force a trial. While it's possible that some of Socrates's opponents might have had genuine religious motives for getting him off the streets, I'm unconvinced. I think it was a setup and that they used the religious angle because otherwise it wouldn't have worked. Naturally, I could be wrong: I'm not suggesting that people haven't killed others or that reigning powers haven't killed people on religious grounds; I'm just saying that in this case I'm not convinced that the religious accusations were anything other than trumped-up charges to call him to account.

---

9. Plato, *Crito* 52b–d.
10. Plato, *Phaedo* 118a.

So what did Socrates die for? Did he die for science, or maybe proto-science? This would be a hard case to make, not because proto-science wasn't already underway (it was), but because Socrates changed the focus of philosophy from the physical world (what many of the pre-Socratics were focused on) to the human individual and society. He wanted to know how we should live. He wasn't uninterested in cosmological enquiries, he just thought the human questions were more important.

More than that, though, Socrates thought that the physical world was not the proper subject of philosophy since we can only ever have opinions about it, not knowledge.[11] You may have heard of his allegory of the cave: a man sits facing a cave wall, on which there are shadows dancing and flickering. That, Socrates thought, was analogous to our perceptions of the physical world: they are constantly changing, there's nothing stable that we can count on staying the same. Then the man stands up, turns around, and sees people dancing in front of a flickering fire; they were producing the shadows. But this still isn't the truly real. The man eventually climbs out of the cave and sees the outside world, light, sky, and sun. The sun represents God and is the height of knowledge. Then the man goes back into the cave, like a missionary, to tell everyone else how to escape to the true reality.[12] And apparently, this required the repeated asking of obnoxious questions.

For Socrates, the physical world was like a pale reflection of what was really real. This is his theory of forms.[13] For example, individual trees are the reflections or shadows of the form of trees. The form is more real and permanent than the individual instantiations of it. We don't need to go into the details of this; the point is that his view was that the physical world is less real than the world of forms, and since the forms are unchanging, we can only ever have knowledge of *them*, not of the physical world. This doesn't mean that Socrates would have been opposed to science at all. He would have used whatever information was available to him. It's just that it's less important than using your mind to get in touch with true reality, i.e., the forms.

This is known as Platonism because it is contained in the writings of Socrates's student Plato. These writings consist of debates between Socrates and numerous people on numerous subjects. It's not clear how much of it actually goes back to Socrates, since Plato probably put his own

11. Plato, *Republic*, bks. 6–7.
12. Plato, *Republic* 514a–520a.
13. Plato, *Phaedo* 78b–84a.

philosophy into Socrates's mouth, especially in the later dialogues.[14] Even so, Plato's philosophy was obviously enormously influenced by Socrates so it's not a different beast altogether. But given the cold shoulder he gives to the physical world, we can't say Socrates was a martyr for science or even proto-science. So what was he a martyr for? At the very least, we can say he was a martyr for free thought, for the right to think for oneself, and we can say that religion was used (perhaps genuinely but perhaps not) to shut him up.

II. HYPATIA

Several centuries later we have another story, and this one definitely involves Christianity. Hypatia of Alexandria (AD 355–415) was a great mathematician and scientist and was on the verge of kick-starting the Scientific Revolution a millennium early. Unfortunately, she was violently murdered by a Christian mob because a) she was an atheist, putting her trust in science instead of God; b) her scientific research and discoveries challenged the Christian faith; and c) she refused to know her place and dared to be an educated woman. About the same time, in the same city, a Christian mob burned down the great Library of Alexandria, destroying much of the collected knowledge of the ancient world. The number of books lost is sometimes estimated to be about half a million. The murder of Hypatia and the destruction of the Library of Alexandria were two of the main factors that led to the fall of the Roman Empire and the thousand-year-long Dark Ages when science and learnin' and the like were spurned until the Scientific Revolution was finally able to get things going again.

This is probably another one of those times where you can already see where this is going. A lot of that information is false. For one thing, Hypatia was not an atheist. She was a Neoplatonist, a philosophy that usually affirms the existence of God. It's *possible* to be a Neoplatonist atheist, but it's like holding the negative sides of two magnets together. Plus, many Christians had an affinity for Neoplatonism. Origen, one of the early church fathers (and who was martyred for his faith), was heavily influenced by the emerging Neoplatonism, having shared the same teacher as Plotinus.[15] Augustine's theology was heavily indebted to Neoplatonism, although he

---

14. Nails and Monoson, "Socrates," pt. 2.
15. Copleston, *History of Philosophy*, 2:27–28; Edwards, "Origen."

moved away from it in his old age.[16] One of the most important Christian theologians in the early Middle Ages, John Scotus Eriugena (ninth century), was a Neoplatonist through and through.[17] The negative theology movement—which basically says you cannot know what God *is*, you can only know what God is *not* (e.g. he is not finite, not limited, not material)—is hugely Neoplatonist in character.[18] And one of the most influential set of texts in the history of Christianity is the work of Pseudo-Dionysius, which situates Christianity within a Neoplatonist framework.[19] It was thought to have been written by Dionysius the Areopagite, who was converted to Christianity in the first century by the apostle Paul (Acts 17:34), but some had their doubts as to its authenticity. It has since been positively shown to have been written long after the historical Dionysius, but this doesn't conflict with the influence it had upon Christianity throughout history.[20] Neoplatonism is a more natural fit (according to me) with pantheism, the idea that God and the universe are identical, but it was still a common enough position within Christianity.

Moreover, Neoplatonism has the same dispassion for the physical world as Platonism does. The physical world is constantly changing, cannot be known, and should not be the main focus of philosophy.[21] This doesn't mean the physical world went unstudied among Neoplatonists or that Hypatia herself did not study it. Neither Platonism nor Neoplatonism *disregard* the physical world, they merely deny that it can be an object of knowledge.

So Hypatia's philosophy was not atheistic and was shared by many prominent and influential Christians. Hypatia herself was admired by many Christians, including many of the most learned Christians of the day. So why do people think she was killed by a Christian mob? Well, the large part of it is because she was killed by a Christian mob. It just wasn't the way the martyr story tells it.[22]

---

16. Copleston, *History of Philosophy*, 2:42–47; Tornau, "Saint Augustine."

17. Copleston, *History of Philosophy*, 2:114, 117; Moran and Guiu, "John Scotus Eriugena."

18. Copleston, *History of Philosophy*, 2:93–95.

19. Copleston, *History of Philosophy*, 2:91–100.

20. Latourette, *Beginnings to 1500*, 210–11; Corrigan and Harrington, "Pseudo-Dionysius the Areopagite."

21. Wildberg, "Neoplatonism."

22. Wessel, *Cyril of Alexandria*, 46–57.

In Alexandria at the time there was a political struggle between rival Christian factions, one led by Cyril, the patriarch of Alexandria, and the other by Orestes, the prefect of Alexandria. Cyril had a hard-line stance against heterodox Christian sects,[23] and had led his followers to retaliate against a Jewish attack on a Christian congregation. Orestes was opposed to all this. Hypatia was a friend and supporter of Orestes, and some of Cyril's people believed she was the person most responsible for preventing any reconciliation between the two.[24]

At some point Orestes was walking through the streets of Alexandria and some monks who were supporters of Cyril were on the sidelines yelling at him. One monk threw a rock at him, and it hit Orestes in the head. Orestes immediately had the monk orested and punished. "Punishment" meant torture, just not torture-to-death torture. Except, oopsie, this time it did: the monk died. Unsurprisingly, the pro-Cyril faction was upset by this. Some of them formed a mob and murdered Orestes's most famous supporter in retaliation. That was Hypatia. It was ugly. The Christian community in general, including the pro-Cyril crowd, was horrified by it.[25]

So was Hypatia killed by a Christian mob? Yes, absolutely. Was it because she was a scientist or a non-Christian or a woman? There is no reason to think these issues had anything to do with it.[26] It was a political act done in retaliation against another Christian faction. So Hypatia wasn't a martyr to science, she was martyred for her political affiliations. It's a horrible story, but it's not a case of Christians killing a scientist for daring to use her knowledge to contradict Christianity, standing up for truth and reality, and refusing to bow her head.[27]

The Library of Alexandria

In some accounts, Hypatia's murderers had not sated their appetite for destruction, so they went to the Library of Alexandria to burn those half a million books which had science and other non-Christian ideas in them. More often, though, the burning of the library is portrayed as an

---

23. Heterodox is neither orthodox nor heretical. It's within the faith but is not the usual take.
24. Chapman, "St. Cyril of Alexandria."
25. Chapman, "St. Cyril of Alexandria."
26. Wessel, *Cyril of Alexandria*, 54.
27. Dzielska, *Hypatia of Alexandria*.

earlier event in Hypatia's life, foreshadowing her own destruction at the hands of Christians.

There's a problem with this though: the Library of Alexandria didn't exist in Hypatia's time. It had been burned down by Julius Caesar's troops in 48 BC, before Jesus' birth and nearly four hundred years before Hypatia's. However, it was rebuilt. And destroyed again, rebuilt, fell into decrepitude, and so on. In its heyday it might have had as many as fifty thousand books, not the half million sometimes touted, but that would still have been enough to make it the largest library in the world at the time. It's just that "the time" in question was not *Hypatia's* time.[28]

The destruction of the Library of Alexandria became associated with Christians because there was an event in AD 391, during Hypatia's life, involving the Serapeum (the temple of Serapis, a Graeco-Egyptian god). This building might have once housed some books, but not at this time. Some workers discovered cult objects from a Mithraic temple on the grounds of the Serapeum and gave some of them to Theophilus, the bishop of Alexandria, and he did the moral and respectful thing: he paraded them through town to hold them up to ridicule (cf. Rom 2:22b). The Serapists[29] understandably took deep offense, but un-understandably they decided to get even by "capturing, torturing, and crucifying" many Christians throughout Alexandria,[30] whether or not they were involved in, or approved of, the taking of the cult objects. In response to *this*, some of the Christian community surrounded the Serapeum. Theophilus offered to let all the Serapists go if they left the cult objects and let the crowd destroy them and the building. They said no. The Serapeum had the high ground. It was located on top of a hill and was easy to defend, so there was a stand-off for a few weeks, with occasional raids by one side or the other. Finally, the Roman emperor, Theodosius Caesar, ordered the people in the Serapeum to vacate it but to take nothing with them. They obeyed his command. Then Roman soldiers, under Caesar's orders, destroyed the cult objects and the building, although you have to figure

---

28. James Hannam, "The Mysterious Fate of the Great Library of Alexandria," http://www.bede.org.uk/library.htm.

29. I guess that would be the right word for the followers of Serapis. Unless they broke off from the main body of Serapis worshipers and started their own temple. Then they would be Serapist separatists. And if they were counselors they'd be Serapist separatist therapists.

30. Dzielska, *Hypatia of Alexandria*, 81.

that some of the mob that had surrounded the place for nearly a month helped out in the Serapeum's destruction.[31]

So, to reiterate: a) it wasn't the Library of Alexandria, it was the Serapeum; b) there were no books or scrolls destroyed, only cult objects; c) it was preceded by violent acts against the Christian community by some Serapists, which were themselves preceded by the denigration of Serapist cult objects by the bishop of Alexandria; d) it was done under orders from Caesar and carried out by Roman troops (for the most part, at least); and e) it took place twenty-five years before Hypatia's death. Thus, linking this event to Hypatia's death, or to the Library of Alexandria, or to the destruction of books and scrolls, or to the fall of the Roman Empire and the beginning of the Dark Ages (which many do) simply doesn't stand up to scrutiny. A film came out in 2009 called *Agora*, which is about Hypatia, and it portrays a lot of these urban legends surrounding her murder, as well as the destruction of scrolls in the Library of Alexandria. The insanely hot Rachel Weisz plays the sixty-year-old Hypatia. It's a good film, it's just one of those "based on a true story except for the main plot points and its overall meaning" movies.

III. BRUNO

In the sixteenth century, Copernicus proposed that the earth revolves around the sun rather than vice versa. He published his ideas right before he died because he was afraid of being mocked by his scientific colleagues. Kepler advocated heliocentrism too, as did Rheticus, Maestlin, Stevin, Digges, and others. Kepler's boss, Tycho Brahe, proposed a modified geocentrism wherein most planets revolve around the sun as the sun revolves around the earth, which is at the center of the universe. None of these people got in trouble for going against the grain on who orbits who.[32]

Then along came Giordano Bruno. He advocated heliocentrism too, but he was arrested by the Catholic Church, imprisoned for years, and finally executed, with all his books being banned. In addition to heliocentrism, he also argued for the plurality of worlds, the belief that all the stars are suns, surrounded by their own planets, with creatures living on them, as well as that the universe is infinite and unbounded, contrary to

---

31. Watts, *City and School in Late Antique Athens and Alexandria*, 190–91.
32. Sorry: "whom."

Ptolemy's cosmology, which says the sphere of the fixed stars is as far as the universe goes.[33]

The question is, if it was heliocentrism that got Bruno in trouble, why did none of the other heliocentrists get in trouble? Well, maybe it was because of his additional claims of the unbounded universe and the plurality of worlds. But other people accepted those too, like Nicholas of Cusa, a Catholic cardinal in the fifteenth century, and none of them got in any trouble. Perhaps it was all three together that did it? Nope, nope, and nope. Thomas Digges believed all three and was fine.[34] So what was it?

Bruno was a Hermeticist. This was a religious tradition based on the *Corpus Hermeticum*, writings ascribed to Hermes Trismegistus from around 1400 BC that influenced many Christians and Muslims.[35] During the Renaissance, when Bruno lived, Hermeticism was somewhat popular, but the texts were then shown to not only postdate Judaism but Christianity as well, having been composed between the second and fourth centuries AD, and its popularity dropped.[36] With regards to Bruno, some scholars argue that he wasn't even a scientist, and was only using heliocentrism, the plurality of worlds, and the unbounded universe to promote his hermetical theological views.[37] However, Kepler treated him as a fellow scientist, and I think we can defer to his judgment. Saying Bruno was a scientist, however, does not imply that he was any good at it: according to Koyré, he was a bad philosopher and a worse scientist.[38]

Part of the issue here is that magic and science both came into their own together and had yet to be fully distinguished from each other.[39] The Scientific Revolution involved just as much magic as science—contrary to popular opinion, magic and witchcraft were much more of an issue in the early modern era than the Middle Ages. Both science and magic were thought to be ways to control nature. The point is that Bruno was more interested in the magic side of the coin than the science side. But since the same type of language was used for both, his writings (as well as others like him) might have had the air of science to later writers.

33. Knox, "Giordano Bruno."
34. Edward Robert Harrison, *Darkness at Night*, 211–17.
35. Yates, *Giordano Bruno and the Hermetic Tradition*.
36. Copenhaver, "Introduction," xl–xlv, l.
37. Shackelford, "Myth 7," in Numbers, *Galileo Goes to Jail*, 62.
38. Koyré, *From the Closed World to the Infinite Universe*, 54.
39. See, e.g., Thorndike's eight-volume work *History of Magic and Experimental Science*.

OK, so maybe it wasn't just the scientific claims, it was those claims plus Hermeticism that got Bruno in trouble. That sounds right—except for the no. The Catholic Church tolerated Hermeticism, as well as the scientific claims. Bruno's advocacy of all these would not have constituted a reason or motive for the Catholic Church to arrest him. Hermeticism *plus* heliocentrism *plus* the unbounded universe *plus* the plurality of worlds only made Bruno idiosyncratic, not an apostate.[40]

It creates a more serious problem, however, when you insist that Hermeticism is the one true religion, and Judaism and Christianity are flawed bastardizations of it. Hermes Trismegistus was alleged to be a contemporary of Moses, and Bruno suggested that Moses took Trismegistus's teachings, altered them, and that's how you get Judaism, and later Christianity. He called for the church to return to its hermeticist origins by rejecting all these holy writings and doctrines. Instead, the church should advocate that Jesus was *not* God incarnate, he was only an especially talented magician; also that the Holy Spirit is just the soul of the world, that the devil will eventually be saved, and ultimately that God is simply the universe and the universe is God (pantheism). So if God is infinite, the universe must be too—hence the unbounded universe. In fact, Bruno's most famous essay, "On the Infinite Universe and Worlds," which argues for the unbounded universe and the plurality of worlds, is barely about science at all, it's essentially an advocacy of pantheism. The ancient Egyptian focus on the sun made heliocentrism attractive to him. To be sure, he didn't merely derive these positions from his theological claims; he gave arguments for them, such as Lucretius's javelin argument for an infinite universe from *On the Nature of Things*.[41] But his acceptance and promotion of the scientific claims were motivated from his perception that they supported his theological views.[42]

Again, Hermeticism was tolerated by the Catholic Church. What wasn't tolerated was to say—as a Catholic priest, which Bruno was—that Christianity is wrong and should be replaced with the true religion it is allegedly derived from. So Bruno was kicked out of the priesthood and went on the lam. He had to keep moving around because he kept offending people: not by his scientific claims but by his actions. His scientific claims were what kept getting him fresh teaching jobs. But he kept losing

---

40. As noted, Thomas Digges affirmed the three scientific claims. He also references Trismegistus, but he wasn't a full-blown Hermeticist.

41. The argument comes at the end of bk. 1.

42. Which isn't a reason to reject them.

them because he would pick unnecessary fights with his new employers, insult their intelligence, and even publish books or tracts to condemn them publicly.[43] If you wrote a book *today* condemning your employer and calling him an idiot, you'd be fired too.

There were also plenty of social and cultural issues in all this. Bruno lived in the decades following the Reformation when the Western Church split into Protestantism and Catholicism. Because of this split, there was a lot of stress on both sides to define the differences between them, a lot of fear about heretics and apostates, and a lot of pressure on the theological powers to do something about it. The whole idea of returning Christianity to its original form was one of the Protestant ideas; they just didn't think that meant returning to a pre-Mosaic Egyptian pantheist who didn't actually exist. So in a sense, Bruno, a Catholic priest, was taking a Protestant idea and taking it *much* further than the Protestants did. The Inquisition was still going strong at this time too, including the Roman Inquisition, the Portuguese Inquisition, and (unexpectedly) the Spanish Inquisition. It was the first of these that had a problem with Bruno; all he had to do to avoid trouble with the Inquisition was stay out of Italy. Pretty easy. He'd get in trouble only if he returned to Italy, so if he didn't go to Italy, he wouldn't get in trouble. Simple. Just don't go back to Italy.

So anyway, Bruno went back to Italy. A fan in Venice was impressed with his writings and offered him a job, which Bruno accepted. Once again, he picked a fight with his employer, and the employer turned him in to the Inquisition. They imprisoned him and told him to recant in full. He refused and stayed in prison for several years, during which he was asked to repent again and again. He refused again and again. Finally, they read a long list of his views, which might have included the unbounded universe and plurality of worlds, and told him to recant of them. He still refused. So they burned him at the stake for heresy.[44]

So was Bruno a martyr to science? He's often portrayed this way, but few historians would agree. He was condemned and executed for his *theological* views, not his *scientific* views. Please don't take this as excusing his execution: obviously he shouldn't have been put to death or imprisoned. I'm just saying the *reason* for his unjust execution was his theology, not his science. "But wait!" I hear you cry, "They may have listed the plurality of worlds and the unbounded universe as two of the claims he needed

---

43. Knox, "Giordano Bruno."
44. Knox, "Giordano Bruno."

to repent of." Here I'll make two points: First, if they did include these claims, it would have been because they were listing off all the things he had written in his books that were unconventional to make the case against him sound as bad as possible. Second, any issue they might have genuinely had with these ideas was Bruno's *use* of them in defending his pantheistic theology. Otherwise, they would have had just as much of a problem with other people who affirmed these ideas, and they didn't.

If he wasn't a martyr to science, was he a martyr to something else? Yes, certainly. He was a martyr to free thought; but it was free thought regarding *religious* ideas, not *scientific* ideas. The reason some people mistake it for the latter is because he had a patina of science over his religious claims. So, like Socrates, he was a martyr for the right to think, argue, and disagree about religion. *Unlike* Socrates, Bruno had something of a superiority complex. Christianity considers Socrates to be a "righteous pagan"; Bruno, not so much.

Chapter 7

Galileo Figaro Magnifico

NEXT UP ARE PEOPLE who were supposedly *persecuted* for their scientific discoveries but not to the point of being killed. There are a lot of urban legends here too. In chapter 5 I mentioned Andreas Vesalius who allegedly got in trouble for dissecting dead bodies—except the text saying he got in trouble is fraudulent, and it doesn't even say that he dissected a dead body, but that he accidentally converted a live body into a dead body via dissection, which would have been awkward.[1] Another example is Roger Bacon who is sometimes portrayed as a Renaissance man born too early. Allegedly, he was persecuted (specifically, imprisoned) for his scientific claims. However, most historians deny he was ever imprisoned at all, since the first mentions of it are late, and those historians who are willing to entertain the possibility say that his imprisonment would have had nothing to do with his scientific views.[2] It would probably have been about his general eschatology (doctrine of end-time), informed as it was by astrology.

But of course, Roger Bacon is not the primary name here. That honor goes to Galileo Galilei. As with some of the other issues I've discussed, the Galileo affair has plenty of truth in it, but also has plenty of urban legend mixed in. Stephen Hawking once compared himself to Galileo as follows:

---

1. O'Malley, *Andreas Vesalius of Brussels*, 304–5.

2. Hannam, *God's Philosophers*, 144–45; Michael H. Shank, "Myth 2," in Numbers, *Galileo Goes to Jail*, 19–27 (esp. 21).

> Throughout the 1970s I had been mainly studying black holes, but in 1981 my interest in questions about the origin and fate of the universe was reawakened when I attended a conference on cosmology organized by the Jesuits in the Vatican. The Catholic Church had made a bad mistake with Galileo when it tried to lay down the law on a question of science, declaring that the sun went round the earth. Now, centuries later, it had decided to invite a number of experts to advise it on cosmology. At the end of the conference the participants were granted an audience with the Pope. He told us that it was all right to study the evolution of the universe after the big bang, but we should not inquire into the big bang itself because that was the moment of Creation and therefore the work of God. I was glad then that he did not know the subject of the talk I had just given at the conference—the possibility that space-time was finite but had no boundary, which means that it had no beginning, no moment of Creation. I had no desire to share the fate of Galileo, with whom I feel a strong sense of identity, partly because of the coincidence of having been born exactly 300 years after his death![3]

Hawking was just making a joke, but the idea that the church tortured Galileo (and perhaps might have executed him if he hadn't recanted) is often used as the exemplar of the antipathy that supposedly exists between science and religion. This gives some people a motive for insisting it's true in order to maintain their belief in the antipathy.

I. WHAT GALILEO DIDN'T DO

Supposedly, Galileo invented the telescope in 1609 and used it to look at the heavenly objects. He saw flaws on the moon and sunspots, thus demonstrating that the heavens are not perfect and incorruptible. He also observed the phases of Venus and discovered four of Jupiter's moons, thus demonstrating that there are heavenly objects that don't orbit the earth. He's also alleged to have shown that the Aristotelian view of gravity, that heavier objects fall faster than lighter ones, was false by dropping rocks of different weights off the Leaning Tower of Pisa. This paved the way for Isaac Newton's law of gravity. Notably, he also affirmed heliocentrism and was imprisoned and tortured by the Inquisition until he disavowed it, since it clearly conflicted with what the Catholic Church had determined to be the case. However, when he finally submitted to the tortures and

---

3. Hawking, *Brief History of Time*, 122.

acknowledged that the earth is immobile at the center of the universe, he quietly whispered, "Eppur si muove" (Yet it does move) under his breath. That's the story.

In point of fact, Galileo *improved* the telescope, but he didn't invent it. We don't really know who did. Hans Lipperhey filed a patent for it in 1608 (and didn't get it), although there were probably precursors. Leonard Digges, Thomas's father, might have invented a rudimentary telescope sometime before his death in 1559.[4] But some have even suggested that Roger Bacon had invented a telescope three hundred years earlier in the thirteenth century. In 1551, Robert Recorde wrote:

> Great talk there is of a glass that he made in Oxford, in which men might see things that were done in other places, and that was judged to be done by the power of evil spirits. But I know the reason of it is good and natural and is wrought by geometry (since perspective is a part of it) and that it stands within reason as much as seeing your face in common glass.[5]

That sounds *something* like a telescope, involving glass, geometry, and perspective, but you have to read a lot into it. And that text was written three centuries after Bacon's time. Another three centuries later, G. W. F. Hegel claimed that Bacon "invented gunpowder, mirrors, telescopes," referencing his contemporary Wilhelm Tennemann's multivolume *Geschichte der Philosophie*,[6] but obviously that's even further removed. Plus, I don't see where Tennemann refers to telescopes (*Ferngläser*). And in case it needs to be said, Bacon didn't invent gunpowder or mirrors either.

Nor was Galileo the first to look at the moon through a telescope, Thomas Harriot (1560–1621) was. He almost certainly didn't drop stones from the Leaning Tower of Pisa, although it's possible someone else might have. However, Galileo *did* discover the phases of Venus and the four largest moons of Jupiter (Io, Europa, Ganymede, and Callisto), which are called the Galilean moons.[7] This was a monumental discovery, however

---

4. Van Helden, "Invention of the Telescope," 13–15, 20–21, 29–30; Ronan, "Origins of the Reflecting Telescope," 337–39.

5. Van Helden, "Invention of the Telescope," 29; Hannam, *God's Philosophers*, 147. I've modernized the language to make it more accessible.

6. Hegel, *Lectures on the History of Philosophy*, 93 (sec. 2, B6c); Tennemann, *Geschichte der Philosophie*, 824–29.

7. How crazy is it that *Galileo* would just happen to be the guy to discover the *Galilean* moons? I mean, the odds against that must be . . . wait, wait, my wife is telling me I'm an idiot. Never mind.

it must be remembered that Tycho Brahe's modified geocentrism, which was very popular, already posited that there are objects that orbit something other than the earth. These discoveries don't, by themselves, show that geocentrism is false or that heliocentrism is true. Galileo also discovered irregularities on the moon and sunspots, which challenged the Ptolemaic cosmology by showing that the heavens are not perfect (by virtue of being composed of the quintessence).

II. THE CENTER OF IT ALL

Although "alternative astronomical hypotheses were freely discussed" at the time,[8] Galileo feared coming out of the closet as a heliocentrist. But his fear, like Copernicus's, was of his fellow scientists, not the church.[9] Indeed, Galileo "was feted by cardinals, received by Pope Paul V and befriended by the future Pope Urban VIII who, in 1620, wrote an ode in his honor."[10] But among scientists, heliocentrism was seen as an ancient idea that had been decisively refuted. In its defense Galileo proposed two theses: a theory of the tides that attributed them to sloshing (and which had them happening only every twenty-four hours instead of every twelve hours), and a theory of shooting stars that treated them as optical illusions. Both of these suggestions were highly speculative and, as it turns out, false.[11]

Nevertheless, heliocentrism was becoming more and more controversial because it went against the age-old scientific teaching about the nature of the cosmos, and this teaching had been unofficially endorsed by the Catholic Church when it endorsed Aquinas's theology, which was heavily indebted to Aristotle. The imprisonment and execution of Bruno was being perceived more and more as having more to do with heliocentrism than with his theological claims (by some, at least). Moreover, as mentioned in chapter 3, *the evidence was inconclusive.* There were certainly enormous problems with Ptolemaic cosmology, but its falsity did not establish heliocentrism. At most, it just meant that Ptolemy could not be accepted in his entirety. To affirm heliocentrism at that time would be to jump ahead of the evidence. This was still true after Galileo's death:

---

8. Sampson, *Six Modern Myths*, 37.
9. Sampson, *Six Modern Myths*, 36–37.
10. Sampson, *Six Modern Myths*, 37.
11. Aiton, "Galileo's Theory of the Tides"; Sampson, *Six Modern Myths*, 38.

geocentrism had stronger *scientific* support than heliocentrism into the second half of the seventeenth century.[12]

Also, as noted in chapter 3, some theologians had tried to read geocentrism into some Bible passages over the centuries. In Catholicism, only members of the church hierarchy were allowed to interpret the Bible. The laity did not have that right. This was one of the main issues that sparked the Protestant Reformation. The Protestants said the Bible is not written in a secret code that only the priestly class has access to, it's written in *language* and we can read it and understand it for ourselves. Naturally, there will be plenty of parts of the Bible that are difficult to understand and to which we can arrive at little consensus. But this is true of language in general. I'm not trying to defend this position (although I do, in fact, believe it), I'm just presenting the motive behind the Protestant doctrine of *sola Scriptura* (Scripture alone). Incidentally, this is *not* the claim that God reveals himself to us only via the Bible,[13] it's the claim that the information necessary to understand the Bible is available to everyone, not only the priestly class. It may not be easy to acquire this information, but its acquisition is not (according to the Reformers) dependent on being a higher-up in the Catholic Church's leadership structure.

III. THE FIRST TRIAL

So why is this relevant? Because Galileo, a Catholic layman, wrote a letter to the Grand Duchess Christina, the matriarch of the family that was sponsoring him, arguing against these biblical interpretations that read geocentrism into the Bible. The passages in question, he argued, are just common phrases that everyone uses and should not be taken as making any scientific claims.[14] Letters were often made publicly available at this time, so his claims became more widely known. These biblical interpretations did not describe the official doctrine of the church, but they had been made by the appropriate people, namely, members of the church hierarchy. So it was a case of a layman telling the priests that they had erred in their interpretation of Scripture, as if he could correct them. This was less than a hundred years after the start of the Reformation, so they were a little peeved by this.

---

12. Graney, *Setting Aside All Authority*.
13. See below, 189.
14. Galilei, "Letter to the Grand Duchess Christina."

In 1615, the Inquisition launched an investigation into the compatibility of heliocentrism with the Bible after a complaint was lodged against Paolo Foscarini, who had argued in favor of their compatibility.[15] By this point, Galileo had become something of a celebrity. The investigation took a year, but Galileo was not summoned. So all he had to do was sit back and let the problem be resolved. But he decided to go to the trial anyway, in December 1615, to defend heliocentrism before the Inquisition, effectively forcing them to pick a side. It would be a great PR move.

Except . . . it wasn't. The Inquisition's ruling was not in accord with Galileo's expectations. First off, in March 1616, they banned Foscarini's book and suspended Copernicus's *On the Revolutions of the Heavenly Spheres*, which had been in circulation without controversy for over seventy years, until they changed it to say that heliocentrism just involves the appearances, not reality (i.e., saving the appearances).[16] The Inquisition ruled that heliocentrism had to be presented this way, so the Index of Prohibited Books, established in 1559 as part of the Counter-Reformation, "published a decree, without mentioning Galileo, that declared that the earth's motion was physically false and contradicted Scripture."[17]

The person in charge of the first trial was Cardinal Bellarmine. This is the introduction of the bad guy to Galileo's good-guy persona, fueled by the fact that he had been in charge of Bruno's case as well. But as always, reality is more complicated than our simplistic take on it: "As for Robert Bellarmine, historians agree on three things. He was in possession of a brilliant intellect, he was a religious fundamentalist and you could not hope to meet a nicer bloke."[18] Bellarmine said he would concede that the Bible passages suggesting that the earth is immobile were figures of speech if it was proven that it moves, but such proof had not yet been discovered—although he did write to Foscarini that heliocentrism made "excellent good sense."[19]

---

15. William R. Shea, "Galileo and the Church," in Lindberg and Numbers, *God and Nature*, 114–35 (esp. 120–21); McMullin, "Church's Ban on Copernicanism, 1616," 170; Hannam, *God's Philosophers*, 315–16.

16. See above, 48–49. Ultimately, they only made ten amendments to it to the effect that heliocentrism was a mathematical hypothesis not an actual description of the universe. The changes were applied via an insert to the books.

17. Maurice A. Finocchiaro, "Myth 8," in Numbers, *Galileo Goes to Jail*, 68–78 (quotation from 70).

18. Hannam, *God's Philosophers*, 309.

19. Sampson, *Six Modern Myths*, 38.

Bellarmine told Galileo that he could not defend or even believe heliocentrism. How you can just decide to not believe something, I have no idea. I couldn't stop believing that I had eggs for breakfast even if you held a gun to my head. However, Galileo could still *discuss* heliocentrism as long as he didn't *defend* it: a fine point, maybe, but important.[20]

IV. THE SECOND TRIAL

In 1623 Cardinal Maffeo Barberini became Pope Urban VIII. He was not only a supporter of Galileo, he was a personal friend, and had written a poem praising him. He was not a heliocentrist himself, but he was opposed to the 1616 ruling against it. And in a completely unrelated turn of events, after becoming pope, he began to be accused of being a milquetoast who was too lenient on heterodox movements and didn't defend the church as rigorously as he should. He even let Tommaso Campanella open a school in Rome to teach blatant heresies in exchange for a personal favor.[21] Urban VIII needed a cause to come down on to appease his critics. As I say, completely unrelated.

Galileo requested and received permission from the pope to write a book presenting the cases for geo- and heliocentrism and the cases against.[22] Since he was allowed to discuss heliocentrism this should have been fine. The pope even asked his good friend Galileo to include one particular point, and Galileo told his good friend the pope that he would do so. So in 1632 Galileo published his *Dialogue Concerning the Two Chief World Systems*. He arranged the book as a series of fictional dialogues or debates. This was and is a common form of academic writing, inspired by Plato's dialogues.[23] In a completely unrelated turn of events, this was around the same time that the criticism of the pope reached its (hitherto) apex.[24]

One of Galileo's characters in the *Dialogue* represented the Aristotelian/Ptolemaic cosmology and the geocentrism that went along with

---

20. Koestler, *Sleepwalkers*, 463.

21. Hannam, *God's Philosophers*, 320–21.

22. Shea, "Galileo and the Church," in Lindberg and Numbers, *God and Nature*, 128.

23. I actually have my students write their term papers in this format. It's more fun. Or less agonizing, at least.

24. Dandelet, *Spanish Rome*, 192.

it. This was the character who repeated the pope's argument.[25] The character's name was Simplicio, possibly named after Simplicius of Cilicia, a commentator on Aristotle. The problem is that Galileo didn't write the *Dialogue* in Latin; he wrote it in common Italian. And in seventeenth-century Italian, *simplicio* meant simpleton, someone who was simple-minded. Galileo, a Catholic layman, called all the geocentrists, including the freaking pope, "Moron." That tends not to go over so well.

He was brought to trial in 1633. Galileo was shocked at the negative reaction: he insisted he was not defending heliocentrism since the arguments he gave in its favor were not decisive. He thought he had just given each side its due. It wasn't his fault that the case for geocentrism was so bad and the pope was such an idiot. (Remember, at this time the scientific evidence supported geocentrism over heliocentrism.)[26] During the trial, a reference to an unsigned (and misplaced) injunction was presented that had supposedly been given to Galileo after the first trial, to the effect that he could not discuss the earth's motion in *any* way. The injunction itself has never been found, there was only a reference to it in Galileo's file that had no signature and so it couldn't be traced back to anybody. Most historians dispute its authenticity. It does seem odd, since Galileo had the blessing of the pope to write the *Dialogue*. He denied ever receiving the injunction and produced the certificate from Cardinal Bellarmine that merely said he could not defend or believe it.[27]

Galileo decided to plea-bargain: he admitted the book could be *interpreted* as defending heliocentrism, but he was surprised since he had not intended it that way. Galileo's surprise was understandable. The only way anyone could possibly misinterpret his book as defending heliocentrism is if *they actually read the book and saw that it was clearly defending heliocentrism.*

Unfortunately, the Inquisition—partially at the instigation of the slighted pope who, in a completely unrelated turn of events, needed a cause to come down on to appease his critics, and who had just been publicly derided as a dimwit by someone he had thought was a friend—decided to make an example out of Galileo. This also had the effect of emphasizing that it didn't matter if you were a celebrity; *they*, the Catholic Church, were in charge. They accused him of vehement suspicion of

---

25. Shea, "Galileo and the Church," in Lindberg and Numbers, *God and Nature*, 130–31.
26. Graney, *Setting Aside All Authority*.
27. Finocchiaro, "Myth 8," in Numbers, *Galileo Goes to Jail*, 70–71.

heresy (an intermediate-level heresy), a punishment far worse than Galileo expected, for discussing heliocentrism in contravention of that unsigned, mislaid, and probably nonexistent injunction. The heresies were the denial that the Bible is authoritative in matters of science and that the earth moved.[28] To be fair, the earth's motionlessness was not made ex cathedra, which would have rendered it infallible and un-overturnable—and, in fact, it *was* overturned in the early nineteenth century when Pope Gregory XVI ensured that a book claiming that heliocentrism had been proven received an imprimatur and removed the *Dialogue* from the Index of Prohibited Books.[29] But back in the seventeenth century, Galileo was consigned to house arrest, and the church heavily publicized the sentencing document, which said Galileo had been "rigorously examined." This was a technical term meaning "tortured."

V. IMPRISONED AND TORTURED

So they imprisoned Galileo during the trial, and afterwards sentenced him to house arrest. Clearly, he was imprisoned. And they said that they tortured him since that's what "rigorously examined" means. Clearly, he was tortured. Right?

Imprisoned

When Galileo arrived in Rome for the second trial on February 23, 1633, he was allowed to stay at the Tuscan embassy as long as he didn't leave until the trial started. This was standard operating procedure; we may not *like* it, we may even *disapprove* of it, but they did this to everyone. The preliminary interrogation took place between April 12 and 30, during which time he stayed at the Inquisition palace. They did have prison facilities there, but they set up Galileo in the prosecutor's six-room apartment and let him have a servant. And just in case you're having similar thoughts to my own ("That's more space than I have in my house with a family of four! And I don't have a servant! My kids won't even do their chores!"), he was not allowed to leave the palace. After the first interrogation, he returned to the Tuscan embassy.

---

28. Finocchiaro, "Myth 8," in Numbers, *Galileo Goes to Jail*, 71.
29. Gerard, "Galileo Galilei."

The rigorous examination took place from June 20 to 24, during which he had to stay in the palace again. We don't know for sure how he was accommodated during this time, so it's possible he was put in a prison cell. That was the custom, after all, for people being investigated but who had not yet had a ruling for them. Again, we don't have to like this, but that's how it was done. However, most historians say he was almost certainly set up in the same six-room apartment as before. So *probably* he was not imprisoned.[30]

After the rigorous examination, he stayed for a few days in the grand duke of Tuscany's palace in Rome. On June 30 he was allowed to go to Siena to live under house arrest at the home of the archbishop of Siena, a close friend whom he *hadn't* held up to public ridicule. Finally, in December, he returned to his own estate near Florence to live under house arrest, where "house arrest" meant he couldn't travel more than a mile or so from his villa. Galileo, at this time, was seventy years old and sick. He was allowed to transfer to another of his estates to be closer to his doctors in 1638. He died in 1642.

Really, in all of this, Galileo was given *extraordinarily* generous treatment. Despite their desire to show people that they didn't play favorites and celebrities were treated no differently than anyone else, they treated Galileo very differently than the common folk. It sounds harsh to us because we live in an "innocent until proven guilty" culture, but for the standards of the day, they were insanely accommodative to Galileo.

So was Galileo imprisoned? If you want to say being forced to stay temporarily in a palace with all your needs taken care of "imprisoned," then sure. How about the house arrest? Again, you could call that imprisonment if you want, so long as you acknowledge that he was free to move around within a mile's radius of his estate. At the very least, his freedom of movement was significantly curtailed, which fits with many people's concept of imprisonment. The usual accusation along these lines, however, is that Galileo was kept in the Inquisition prison, against which there is strong evidence and for which there is none. If that's what we mean by imprisonment, then Galileo was not imprisoned. However, everyone *thought* that he had been kept in a prison cell at the Inquisition palace (since that and wearing an onion on your belt was the style at the time) and the Catholic Church didn't correct it. It wasn't until the late eighteenth century that the imprisonment claim was disproven, but it's had a hard time making its way into the public consciousness.

30. Finocchiaro, "Myth 8," in Numbers, *Galileo Goes to Jail*, 73–74.

Tortured

The case *for* Galileo being tortured seems pretty solid: the people who would have tortured him said that they tortured him. Unfortunately (or fortunately, depending on your point of view), the case *against* it is overwhelming. This wasn't proven until the late nineteenth century, and it's had an even harder time making its way into the public consciousness.[31]

1. Torture was only ever done for formal heresy. Galileo was not accused of formal heresy.
2. The various Inquisitions went through stages when torturing was more common or less. At the time of Galileo's trial, the Inquisition authorities in Rome rarely tortured anyone.
3. Inquisition rules specifically exempted the old and the sick from any kind of torture. In 1633, Galileo was both.
4. The tortures that were used would have permanently injured Galileo. He was not permanently injured by anything during the trial.
5. The tortures that were used would have rendered Galileo physically incapable of attending the sentencing the day after they would have taken place. Yet Galileo attended the sentencing.
6. Inquisition rules spared clerics (the lowest rung of the Catholic hierarchy) from being tortured. Galileo had become a cleric in April 1631 to receive an ecclesiastical pension.
7. There had to be a formal vote by the Inquisition consultants authorizing the torture. No such vote is recorded.
8. Torture sessions had to be meticulously recorded, including the victim's screams and moans (fun!). There is no such record in Galileo's case.
9. Torture sessions had to wait ten hours after the victim's latest meal before the torture party. The pace of Galileo's trial did not leave any gap long enough for this to take place.
10. Confessions under torture had to be ratified twenty-four hours later to ensure that the victim was not just saying what the torturers wanted to hear in order to stop the torture. This is because they

---

31. The following points are taken from Finocchiaro, "Myth 8," in Numbers, *Galileo Goes to Jail*.

didn't use torture as *punishment* per se, but to ascertain the truth. There is no such ratification in Galileo's case.

For these reasons, nearly all historians (if not all) maintain that Galileo was never tortured. The Catholic Church *claimed* that he was "rigorously examined" to send a message, but it wasn't true. This was really an own goal by Christianity. They said they tortured him and let the story of his imprisonment spread throughout society to make a point about who is in charge.

There is one more detail that the crowd who believe Galileo was tortured makes. They point out that the inquisitors threatened to torture him and showed him the torture instruments they would use. In the minutes of the second trial, the inquisitors say (speaking to Galileo in the third person), "And he was told to tell the truth, otherwise one would have recourse to torture."[32] This sounds like it defeats a lot of the points made above.

The first response to this is that it doesn't say he was tortured or even that he was shown any instruments of torture. It says he was *threatened* with torture—but it wasn't if he continued to defend heliocentrism, it was if he had committed perjury by lying under oath. Again, torture was not used by them as a punishment but as a way of determining the truth. If someone says something under oath, but there are good reasons to think they are lying, the best way (so they thought) of finding out if they were lying was to torture them to the point where telling inconvenient truths would be less traumatic—and then they had to confirm later that what they said under torture was true and not just said to make the torture stop.

The second response is that "one would have recourse to torture" was said to anyone who was making a pronouncement under oath that the Inquisition had reason to believe was not true. And in this case, they had such reason: Galileo clearly believed heliocentrism. He had sworn under oath that he did not, but he was obviously lying, so they said the standard line that he had said this under oath and if it turned out he was lying, they would supposedly resort to another form of determining the truth. Except they *wouldn't* have tortured him because of the many and sundry reasons given above. Since Galileo was not a candidate for torture, it was simply a legal formality.[33] Essentially, it was their horrific version of "Do you solemnly swear to tell the truth, the whole truth,

32. Finocchiaro, *Trial of Galileo*, 134.
33. Finocchiaro, *Galileo Affair*, 38.

and nothing but the truth, so help you God?" The Roman Inquisition almost never tortured anyone at this time, and certainly wouldn't have tortured someone who was old, sick, a cleric, and wasn't even convicted of a torture-worthy crime. Saying they would have "recourse to torture" was just a procedural formula.

And in case it needs saying, the claim that Galileo whispered under his breath "And yet it moves" isn't even mentioned until over a century after his death. There's no evidence he actually said it.[34] But of course, there *wouldn't* be, would there? That's the whole point of whispering.

VI. CONCLUSIONS

So the Galileo affair is much more complicated than how it is usually presented. I haven't even mentioned how politics and patronages played into it. Galileo was a court favorite in Tuscany, and there were power struggles involved. But the primary issue was who gets to decide how to interpret the Bible in the aftermath of the Reformation. Additionally, Pope Urban VIII used heliocentrism as an occasion for showing that he wasn't wishy-washy like his critics claimed, as well as to get back at Galileo for insulting him. Moreover, Copernicanism was popular among German Protestants, and this provided the Catholic Church with another motive for dissociating themselves from it and making it a boundary issue to determine who was and wasn't a good Catholic.[35] It was a time when everyone was drawing lines in the sand. For these reasons, the Inquisition wanted to make an example out of Galileo to show that no one—not a celebrity, not a Tuscan court favorite—was above the authority of the Catholic Church. None of this *directly* involved heliocentrism, but it was a perfect fit for addressing the larger concerns. If heliocentrism hadn't served these purposes, other issues would have had to be used. Heliocentrism is the surface level of what was at issue; but scratch that surface and it becomes much more complicated.

And yet . . .

---

34. The earliest account comes from Baretti: "The moment he was felt at liberty, he looked up to the sky and down to the ground, and, stamping with his foot, in a contemplative mood, said, *Eppur si move*; that is, still it moves, meaning the earth" (*Italian Library*, 52).

35. Shea, "Galileo and the Church," in Lindberg and Numbers, *God and Nature*, 128. On boundary issues, see below, 124.

In 1616, after the first trial, the Index of Prohibited Books, an official office of the Catholic Church, published a decree that declared that the earth's motion *contradicted Scripture*. And in 1633, Galileo was charged with intermediate level heresy. What heresies was he charged with? First, that the Bible is not authoritative in matters of science; and second, *that the earth moves*. So as nice as it would be to end on the point that the scientific issues were just a superficial excuse, these two points make that impossible. The Catholic Church took a stand by affirming a false cosmology to be what the Bible teaches, and while this was not made ex cathedra (that is, infallibly), it had a significant negative impact on the history of science. Catholics certainly made enormous contributions to science,[36] and still do, but the Galileo affair discouraged some early scientists, like Descartes, from publishing ideas that might irk the powers that be.

---

36. Lawrence M. Principe, "Myth 11," in Numbers, *Galileo Goes to Jail*, 99–106.

Chapter 8

Evolutionary Myths

IN THIS CHAPTER, WE'LL be looking at numerous urban legends surrounding evolution. We have to eliminate these misunderstandings so we can have a clear look at it. But none of this is a challenge to biological evolution itself. I want to emphasize this: *I accept evolution and am not challenging it*. I'd be shocked if it were not true. Fortunately, whether it's true or false is irrelevant to my Christian faith.

The urban legends we're addressing are about historical issues, not scientific issues. I'm going to focus on the myths that are used to support evolution over against religion, but I'll also deal with a few myths about evolution believed by its critics.

I. SCIENCE VS. RELIGION, REASON VS. EMOTION

Many people have the idea that evolution (and science in general) appeals to our rationality while religion appeals to our emotions, and that explains why evolution was accepted by many and rejected by many. Obviously this is pretty simplistic, so you have to figure when you start adding details the case becomes much less clear.

The beauty of evolution

For one thing, evolution can be and has been taken in a very poetic, artistic, emotionally appealing way. As already mentioned, in the eighteenth and nineteenth centuries, there was a pervasive sense of developmentalism in Western civilization—that idea that science and technology would bring about a man-made utopia. Seeing this same pattern in the biological world seemed very poetic and apropos: this developmental pattern was thought to be present in all levels of reality, a sort of theory of everything, and they expressed it in all levels of society, including art. One example is Richard Wagner's *Ring of the Niebelung*, which consists of four operas that are each about four hours long. The *Vorspiel* or overture to the first opera (*Das Rheingold*) is a microcosm of this sixteen-hour piece of music and its developmentalist spirit. It starts with the contrabasses playing a low E-flat, very quietly. Then the bassoons come in a fifth above them. Slowly, the other instruments of the orchestra come in, developing the piece, making it more complex, more beautiful, richer, and much, *much* louder. Then, just as it reaches its glorious apex, that low E-flat still being played by the contrabasses, . . . some German woman starts singing and ruins the whole thing. Wagner wrote the Ring Cycle between 1848 and 1874, the *Vorspiel* itself being written (in rough draft) in 1853–54, prior to Darwin's publication of *On the Origin of Species*. Wagner wrote a letter to August Rockel outlining the theme of the whole four-opera cycle:

> The development of the whole poem sets forth the necessity of recognising and yielding to the change, the many-sidedness, the multiplicity, the eternal renewing of reality and of life. Wotan rises to the tragic height of willing his own destruction. This is the lesson that we have to learn from the history of mankind: to will what necessity imposes, and ourselves to bring it about.[1]

Natural law is what "necessity imposes," and the developmentalism of the time was seen as a natural law in excelsis, a law that applied to all subjects. Evolution, in biology and elsewhere, beautifully expressed "the eternal renewing of reality and of life."

If you want a more recent example, I'd suggest Stanley Kubrick's film *2001: A Space Odyssey*. Part 1 involves a tribe of hominids being banished from their water hole by another, stronger tribe, condemning the first tribe to death. The first tribe's leader has the idea, for the first

---

1. Wagner, *Richard Wagner's Letters*, 97; "World's Last Night," in C. S. Lewis, *World's Last Night*, 102.

time, of using a tool (a weapon) to take back the water hole. It's presented as the evolutionary step that defines them as human thereafter, as well as animals taking control of their own evolution.[2] As the leader has this idea, Richard Strauss's symphonic poem *Also sprach Zarathustra* plays, which you'd recognize immediately if you heard it. This piece is based on Friedrich Nietzsche's book of the same name (*Thus Spoke Zarathustra*) and the concept of the superman or Übermensch who raises himself up to the stage beyond humanity. The final part of the film culminates with humanity being invited to take the next step forward in evolution, with *Also sprach Zarathustra* playing again. It's brilliant, beautiful, and a clear appeal to emotion.

So science, and evolution in particular, appeals to our emotions *as well as* our rationality. What about religion? Does it appeal only to our emotions? Do people believe it only because they want it to be true? Of course, there are plenty of aspects of religion and Christianity in particular that most people would like to be true: we are loved by the very ground of existence, for example. But there are plenty of aspects that we *don't* want to be true, like the doctrine of hell. The only people who want *that* to be true are those who want it for others whom they especially loathe.

The most obvious example though is something that attaches to all religions as a general concept: if religion is true, *you are not the main character of your own story*. Too many people don't grasp how utterly devastating this realization can be. It's not merely that there are things much more important than you, nor even that you should base your life on these more important things. You can acknowledge this and still be the main character in your own story. But religion requires you to go further.[3] I once heard Christopher Hitchens say in a debate that there is no emotional appeal to atheism, the only motive for accepting it is rational. This blew me away: there are *enormous* motives to want atheism to be true, particularly that it allows you to be the main character in your own story. You get to be your own boss, the king in the castle, the master of your domain. To suggest that this doesn't count as a nonrational motive to want atheism to be true simply isn't serious.

---

2. When Kubrick made this film, it was generally thought that the distinctive aspect of human beings is that we use tools. When we found other animals using tools, we suggested that only human beings *make* tools. Then we found other animals doing that too. Now the suggestion is that we're the only animals that sell tools on eBay.

3. Matt 16:24–25; Rom 6:6; Gal 2:20; 5:24; 6:14.

And really, just how emotionally appealing *is* Christianity in the first place? C. S. Lewis said if it were only a matter of enjoying the storytelling and its contents, he preferred Celtic, Norse, and Greek mythology over Christianity.

> Is Theology merely poetry? . . . Does Christian Theology owe its attraction to its power of arousing and satisfying our imaginations? Are those who believe it mistaking aesthetic enjoyment for intellectual assent, or assenting because they enjoy? Faced with this question, I naturally turn to inspect the believer whom I know best—myself. And the first fact I discover, or seem to discover, is that for me at any rate, if Theology is Poetry, it is not very good poetry.[4]

Sociological motives

In addition to the aesthetic motives for accepting evolution, there were also pretty clear sociological motives as well. Notably, evolution was inspired by Victorian society, which wanted to believe that the more well-off deserved to be so. Evolution says the stronger survive and the weaker perish: in Darwin's culture, the rich survived and the poor perished. This led to a feedback loop where evolution reflected society, and society then tried to reflect evolution.[5] Some rich people used Darwinian evolution to justify themselves. To deny the rich their riches supposedly went against a fundamental law of nature. It was a sort of divine right of kings, except with nature substituted for God and the rich substituted for royalty. The rich were where they were by natural design.[6] This is often called social Darwinism, the idea that some people are more evolved (and thus better) than others. This can be applied intra- or inter-culturally. The first led some to affirm an unrestrained laissez-faire capitalism,[7] while the second led some to affirm colonialism. Eventually social Darwinism led to

---

4. "Is Theology Poetry?," in C. S. Lewis, *Weight of Glory*, 118.
5. Sampson, *Six Modern Myths*, 62–65.
6. This is the idea in the Wagner quote above: we should recognize what nature is bringing about—"what necessity imposes"—and work towards it as well, even if what it's bringing about is our own destruction. This commits the naturalist fallacy. Saying something *is* a certain way doesn't mean that it *should be* that way.
7. I'm not criticizing capitalism, I'm just pointing to how some people used evolution to justify it.

eugenics, according to which society's weaker elements should be forcibly removed from the gene pool by sterilization or even elimination.[8]

None of this calls evolution into question. None of it means evolution is not true, rational, or scientific. It means that *we are human* and, as such, our imaginations and emotions will always play roles in how we form our beliefs. And they *should*: How could science even work if scientists didn't use their imaginations and their sense of wonder?

II. ACCEPTANCE AND REJECTION

Who immediately accepted or rejected evolution?

Many people have a skewed understanding of scientific advancement. They think that when a scientist suggests a new theory that turns out to be correct, it is immediately accepted by the scientific community, who recognize its good sense, while it's rejected by the traditionalists. The same claim is made of Darwinian evolution. Like Copernicus, Darwin delayed publishing his theory. But he tended to write slowly, and wrote only one book at a time, so it's understandable. If there was any fear involved, it would have been like Copernicus's fear of his fellow scientists, not religious folk.[9]

The scientific response to Darwin ran the whole gamut. Some rejected his theory because Darwin didn't have much evidence in its favor. Some accepted it because it had a great deal of explanatory power. Some rejected it because at the time the scientific evidence indicated that the earth was only a few million years old, and this wasn't enough time for evolution to take place. Some accepted it because they thought they could use it to advocate for naturalism and secularism. Some rejected it because they did not view biology (natural history) as a proper science at the time.[10]

The use of it to advance naturalism and secularism made it a discussion piece on the popular level among laypeople more than scientists. Scientific societies tended to ignore Darwinism in the aftermath of the publication of *Origin of Species*. Darwin was given numerous scientific awards, but they were based on his studies on earthworms and barnacles.

---

8. Sampson, *Six Modern Myths*, 65–70.
9. Sampson, *Six Modern Myths*, 57.
10. Sampson, *Six Modern Myths*, 56–60.

In fact, nearly every award Darwin received tried to distance him from evolution as it was considered too speculative and not very scientific.[11]

The corollary of this myth is that religious people supposedly *rejected* Darwin right out of the gate, and for religious reasons. They didn't like the suggestion that human beings are animals. However, the traditional definition of human beings in Western civilization is "rational animals" (Aristotle) or "rational, mortal animals" (Augustine), so this doesn't wash. In any event, the religious response to Darwinian evolution ran the whole gamut too. Some rejected it for theological reasons, like Charles Hodge. Some tried to synthesize evolution with Christianity, like Charles Kingsley. Some withheld judgment until the evidence came in, like C. H. Spurgeon. Some accepted it for scientific reasons, like B. B. Warfield.[12] Some even accepted it for *theological* reasons: interestingly, one group that stood out most distinctly in favor of Darwinian evolution was *evangelical Christians*.[13]

Did Darwin become an atheist because of evolution?

Darwin was a Christian as a young man, but he later rejected it, and some suggest it was because his evolutionary theories made mincemeat of it. Indeed, he *was* a Christian as a young man and he *did* reject it later, but evolution had nothing to do with it. He was struck with two personal tragedies: his father died in 1848 and three years later his beloved daughter Annie died. The first case made him struggle with the problem of hell as his father was not a Christian while the second made him struggle with the problem of evil. Eventually, he gave up his Christian faith.[14] Yet he still donated his money and resources to local churches, kept company with Christians, and all of this was sufficient for him to be buried at Westminster Abbey, which is supposed to be reserved for Christians.[15]

There is debate whether Darwin was an atheist, an agnostic, or a "muddled theist." He wrote in his autobiography that he considered himself an agnostic,[16] but a decent case can be made for the muddled

11. Sampson, *Six Modern Myths*, 61.
12. Sampson, *Six Modern Myths*, 55–56.
13. Livingstone, *Darwin's Forgotten Defenders*.
14. James R. Moore, "Myth 16," in Numbers, *Galileo Goes to Jail*, 142–51 (esp. 146–47).
15. Moore, "Myth 16," in Numbers, *Galileo Goes to Jail*, 150–51.
16. Darwin, *Autobiography*, 94.

theist camp. During the last year of his life, he told people that he was convinced the world is not the result of chance and that sometimes he was overwhelmed with the sense that living creatures must have been created by a mind.[17] But that's as far as we can go.

Some try to go further though: in 1915, Lady Elizabeth Hope claimed she visited Darwin in his final days and he recanted of evolution and accepted Christianity again. She did get some things right in her account, like the description of Darwin's home and some of his habits (although biographies had been published about him by this point so she could have gotten it secondhand). But his family didn't recall her ever visiting and said it would have been enormously out of character. Not to mention that the alleged tensions between evolution and Christianity were only in nascent form when Darwin died (although he was obviously aware of them). Lady Hope applied the issues as they had developed in the early twentieth century to the 1880s and to Darwin himself. Most historians don't take Lady Hope's claims seriously.[18]

III. THE HUXLEY-WILBERFORCE DEBATE

Two specific events played a prominent role in setting Christianity against evolution, one in the United Kingdom and the other in the United States. The UK event was a debate between T. H. Huxley and Samuel Wilberforce in 1860. The story is that Huxley was a scientist who was defending Darwin's theory of evolution while Wilberforce was a churchman—he's almost always referred to as Bishop Wilberforce to emphasize that fact—and had to be coached about evolution in order to participate. Because evolution was seen by all as making good sense and being good science, Huxley was running away with the debate, so Wilberforce decided to switch gears and just flat out insult him, asking him whether it was his grandfather or grandmother who had had carnal relations with an ape in order to produce him. Huxley responded he'd rather be descended from an ape than from a talented, eloquent man who used his gifts to obscure the light of scientific truth. BOOM shakalaka. Science wins, religion loses.

When we add a few more details, though, the picture again becomes less clear. Wilberforce was indeed a bishop, but he was also the

---

17. Darwin, *Life and Letters*, 316 and note; Campbell, "What Is Science?"; Campbell, *What Is Science?*, 62–63.

18. Moore, "Myth 16," in Numbers, *Galileo Goes to Jail*, 148–50.

vice president of the British Association for the Advancement of Science and a Fellow of the Royal Society. He had also written a nearly forty-page review of *Origin of Species* in which he objected to evolution "solely on scientific grounds.... We have no sympathy with those who object to any facts or alleged facts in nature ... because they believe them to contradict what it appears to them is taught in Revelation."[19] Darwin himself thought Wilberforce's critique was very astute,[20] so obviously he did not need to be coached about what Darwin's book said. Also, his nickname was "Soapy Sam," which, unfortunately, referred to his slipperiness, not his cleanliness.

Huxley was a scientist and friend of Darwin who was later nicknamed "Darwin's Bulldog." He was one of the main players in the late nineteenth century seeking to set religion and science against each other in order to promote naturalism and secularism, and he immediately saw the value of Darwin's theory to this end. He's also the one who coined the term "agnostic" for someone who doesn't know or says we can't know whether God exists, from the Greek for "no knowledge" (the Latin form was already taken: "ignoramus"). Obviously, there were plenty of good things and bad things about both Huxley and Wilberforce, like there are about everyone else. I'm just trying to balance out the one-dimensional picture that paints Huxley as the good guy and Wilberforce as the bad guy.

As to the debate—well, it wasn't really a debate. A paper was read on evolution at Oxford and there were multiple respondents, including Huxley, Wilberforce, J. S. Henswell, Benjamin Brodie, Joseph Hooker, and others. Huxley chose not to engage at first because the audience was mostly nonscientists, or at least *nonprofessional* scientists, and he didn't think they could grasp the issues in question. Wilberforce argued along the same lines as his published review that Darwin's theory didn't have sufficient scientific support (which, at the time, was true), and wasn't a proper case of inductive inference (which was a common objection).[21]

Another problem Wilberforce had with evolution was that he believed it implied that certain races or ethnicities of human beings were more evolved than others and thus superior, allowing for the more evolved to oppress the less evolved. This is that social Darwinism I mentioned earlier. Samuel Wilberforce was the son of William Wilberforce, who played such a huge role in the abolition of slavery in Western civilization; and dehumanizing other groups of people was an excellent

19. Wilberforce, Review of *On the Origin of Species*.
20. Brooke, "Wilberforce-Huxley Debate," 139.
21. Jensen, "Return to the Wilberforce-Huxley Debate," 164, 166.

method for justifying their enslavement. Thus, Samuel Wilberforce was concerned that evolution could be used to rationalize slavery as well as racism and white supremacy. The reason we know this was one of his concerns is because *that was what the discussion was about*. The paper the respondents were responding to was by John William Draper, about how evolution related to the intellectual development of European culture. Draper expanded the material into a two-volume work, which begins with mentioning his contribution to the meeting as "the physiological argument . . . respecting the mental progress of Europe," when compared to other ethnic groups. "All over the world physical circumstances control the human race. They make the Australian a savage; incapacitate the negro, who can never invent an alphabet or an arithmetic, and whose theology never passes beyond the stage of sorcery."[22]

Wilberforce's comment to Huxley—if he's descended from an ape, was it his grandfather or grandmother?—was reported in the press and correspondence immediately, and was seen as rude and improper; which shouldn't be surprising since it was. I could be wrong, but I suspect that Wilberforce mistakenly thought Huxley wasn't applying the same standard to himself that he was applying to others and was trying to bring this out. "OK, so you think the fact that there are cultures that are less successful than our own is explained by their degrading evolutionary origins, but according to your thesis, *your* origins are just as degrading, so you're no better than them."

The first account of the exchange did not mention Huxley's response to Wilberforce, that he'd rather be descended from an ape than from a gifted man who didn't care for truth, but it is mentioned soon thereafter in letters by Huxley and J. R. Green, as well as a few media accounts.[23] However, it appears to have gone largely unnoticed and wasn't remembered. Both Huxley and Wilberforce thought they'd won, but then so did Hooker (which is a defensible claim). At least one scientist, Henry Baker Tristram, switched his position because of the discussion . . . to Wilberforce's side.[24]

---

22. Draper, *History of the Intellectual Development of Europe*, 1:iii, 26–27. To be fair, he also describes certain European races as inferior as well.

23. Jensen, "Return to the Wilberforce-Huxley Debate," 168–69; England, "Censoring Huxley and Wilberforce," 376.

24. David N. Livingstone, "Myth 17," in Numbers, *Galileo Goes to Jail*, 152–60 (esp. 156).

The Huxley-Wilberforce debate didn't enter the science vs. religion lexicon until decades later when biographies of Huxley and Darwin were published and referred to it, portraying Huxley's rejoinder as a scientific coup d'état against religion.[25] Draper wrote *The History of the Conflict Between Religion and Science* in 1874, in which he ransacked history looking for anything and everything that could be twisted into a conflict between science and Christianity.[26] Yet he doesn't mention the Huxley-Wilberforce debate. This, despite the fact that he was not only present that day, but the whole discussion had been a response to his own essay. This was only fourteen years afterwards, though, and the story's religious angle hadn't taken shape yet.

IV. THE SCOPES MONKEY TRIAL

The US event that set Christianity against evolution was the Scopes trial, which took place in Tennessee in the summer of 1925. The story here is that Tennessee had just passed a law prohibiting the teaching of evolution in schools. This was fueled by ignorant religious believers (and racist southern hicks at that) who, like always, got in the way of science and progress and truth. Not long after the law took effect, a simple, honest biology teacher named John Scopes in Dayton, Tennessee, was arrested and imprisoned for breaking it by merely doing his job. It was a big battle in the culture war that's been raging ever since.

Interestingly, the good guy in the story isn't Scopes himself but his defense attorney Clarence Darrow, the voice of cool reason. The bad guy was William Jennings Bryan, the prosecutor. Once all the religious nuts had their say about how Scopes was seducing their children away from the light with his evil science, Darrow, in a very unusual move, called Bryan to the stand to ask him exactly what the problem was with evolution. Bryan sermonized about how the universe is only six thousand years old, but when asked a few pointed questions by Darrow, found himself unable to respond to even the simplest of objections. When the smoke cleared, Scopes was convicted but given a slap on the wrist, while Bryan died of embarrassment a few days later for having humiliated himself and Christianity. So it was a Rocky Balboa victory: *technically* the good guys lost, but in actuality they triumphed over the forces of evil and changed

---

25. Livingstone, "Myth 17," in Numbers, *Galileo Goes to Jail*, 155.
26. See below, 191.

the world for the better. A play was written about this glorious victory called *Inherit the Wind*, which was made into a movie starring Spencer Tracy in 1960.

Anyway, the first minor correction to make to this story is that it was all a publicity stunt. Tennessee had passed the law making the teaching of evolution a finable offense. The ACLU asked for volunteers to challenge it and said that they would pay whatever expenses and fines came out of it. School officials and community leaders in Dayton asked Scopes if he'd be willing to be the guinea pig in order to bring some notoriety to their town, which was steadily decreasing in population and influence.[27] Scopes agreed despite the fact that a) he was really the football coach and was only temporarily substituting for the science teacher; b) when he did teach, he taught only mathematics and physics, not biology; and c) he didn't remember if he'd even brought up evolution in these classes (but he agreed to say that he had for the sake of the case). He was never arrested or imprisoned and was never threatened with arrest or imprisonment. Again, the law entailed only a fine, not jail time.[28]

Part of the problem was that the state of Tennessee required schools to use a particular biology textbook, *Civic Biology*, by George William Hunter, which taught evolution. So it was a catch-22: teach evolution and you're breaking the law; don't teach evolution and you're breaking the law (by not using the required textbook).[29]

As to the players: Clarence Darrow was a very clever and very cantankerous antireligion agnostic. He was also a celebrity lawyer, having recently been the defense attorney in the Leopold-Loeb thrill-kill case, which you can look up if you think you've been too cheerful lately. William Jennings Bryan, who had been the Democratic nominee for president of the United States three times, did not accept young-earth creationism. Very few people did at the time. That interpretation was primarily a phenomenon within Seventh-day Adventism, a denomination that began as a cult in the mid-nineteenth century based on the visions of Ellen White. Those visions were the source for the young-earth creationist scenario.[30] It didn't begin to get popular among other Protestants until 1961, when

---

27. Edward J. Larson, "Scopes Trial in History and Legend," in Lindberg and Numbers, *When Science and Christianity Meet*, 245–64 (esp. 247–49).

28. Sampson, *Six Modern Myths*, 54.

29. Larson, "Scopes Trial in History and Legend," in Lindberg and Numbers, *When Science and Christianity Meet*, 248–49.

30. Numbers, *Creationists*, 89–119, 215–16, 222–24.

Henry Morris and John Whitcomb published *The Genesis Flood*. While Bryan accepted what science said about the age of the earth, and even accepted biological evolution in general, he didn't accept its application to the origin of humanity.[31] The most popular Christian take on the Bible and the age of the earth at the time was the gap theory, which held that the biblical account of creation includes a gap of unspecified length before the events get going. Bryan, however, held to the also-popular day-age theory, that the days of creation are unspecified periods of time.[32]

Darrow had a history with Bryan due to their disagreement on religion, despite their agreement on politics. As soon as he found out Bryan was going to be the prosecuting attorney Darrow tried to ingratiate himself into the proceedings. The ACLU didn't want him: they already had their own lawyers in place to defend Scopes, and Darrow's presence wouldn't let them fully control the media circus they were orchestrating. So Darrow went to Scopes directly and offered his services pro bono. Scopes accepted, and the ACLU reluctantly allowed him to play a role, although he still wasn't the lead defense attorney.[33]

Bryan's prosecuting team had refused to get into the larger issues about science and religion and had focused on whether Scopes had broken the new law or not. Partially because of this, the defense lawyers, including Darrow, called Bryan to the stand to defend his take on Christianity. When Darrow questioned Bryan, he just started listing off some of the problems he had with Christianity and the Bible and challenged Bryan to respond to them. Bryan's responses were relatively weak, but he insisted he wouldn't run from anyone challenging his faith. There was a huge audience and the Dayton courthouse was in disrepair, so they had to move everything outside to accommodate all the people who wanted to watch (it was also being broadcast on radio live all over the country). After several hours in the hot sun, the judge halted the proceedings until the next day, but then never started them up again. Bryan had planned an extensive closing statement to deal with everything—including his concerns about eugenics—but the defense thwarted him by declining to give a closing statement themselves, and so by law Bryan couldn't give

---

31. Ronald L. Numbers, "Creationists," in Lindberg and Numbers, *God and Nature*, 391–432 (esp. 402–3).

32. For a contemporary defense of the gap theory, see Custance, *Without Form and Void*. For the day-age interpretation, see Ross, *Matter of Days*.

33. Larson, "Scopes Trial in History and Legend," in Lindberg and Numbers, *When Science and Christianity Meet*, 254.

his remarks either. It was agreed on all sides that Scopes had broken the new law (or at least that he *claimed* he had broken the new law, since he didn't remember if he had mentioned evolution in class), so he was fined one hundred dollars, which the ACLU paid per their agreement. Bryan did die a few days later, but it wasn't from embarrassment since dying of embarrassment isn't actually a thing. It was probably due to being an old man who had just spent an extended stressful period of time in the heat.[34]

None of this explains why Tennessee would even pass a law forbidding the teaching of evolution, or why Bryan was opposed to evolution despite accepting everything else that contemporary science was saying. Yes, it was a publicity stunt, but for a publicity stunt to *work* there must be public *interest* in the issue. So why was evolution an issue? One factor was the rise of fundamentalism. *The Fundamentals*, consisting of ninety essays, were published between 1910 and 1915 and responded to current concerns by specifying and defending the core tenets of Protestant Christianity. While *The Fundamentals* did not take issue with the age of the earth and universe or other scientific discoveries, they did single out evolution for opprobrium.[35] Unfortunately, as is often the case, the specifying and defending of what is supposedly necessary to be a Christian led many people to use those specifications as boundary issues to determine who's on the inside and who isn't. Specifying boundaries like this is necessary and everybody does it to some extent, but it's part of human nature to oversimplify things, and this can be disastrous when it comes to boundary issues.[36] Anyway, the point is that some people at the time of the Scopes trial took *The Fundamentals* as a method of judging who was and wasn't a Christian—or at least the "right kind" of Christian.

But this just pushes it one step back: Why were *The Fundamentals* opposed to evolution? The simple answer is because they thought it was inconsistent with Christian teachings about the origin of humanity. But we've already seen that the Christian response to evolution was all across the board with plenty of people—specifically evangelical Christians, who one would think would be most prone to agree with *The*

---

34. Larson, "Scopes Trial in History and Legend," in Lindberg and Numbers, *When Science and Christianity Meet*, 255–61; Numbers, "Creationists," in Lindberg and Numbers, *God and Nature*, 403.

35. Orr, "Science and Christian Faith"; "Early Narratives of Genesis," esp. 6:93–94, 96).

36. E.g., Dunn, *Romans 1–8*, lxvii–lxxii.

Fundamentals—coming out *in favor* of Darwin's theory, at least initially.[37] So what accounted for the change? Some of it, no doubt, was because of the efforts of Huxley and others to use evolution as a wedge to separate religion and science as mutually exclusive, which was an idea that had taken a greater hold on culture over several decades. But that wasn't the whole thing.

To complete the picture, we have to bring in our old friend *social Darwinism*. If evolution applied to the world of biology, then a similar case could be made (supposedly) for applying it to the worlds of sociology and anthropology. This entailed (supposedly) that certain races of people are superior, more evolved than others, and (supposedly) white people and Western civilization are at the top of that particular dung heap we call humanity. This led many countries to enact eugenics programs to keep retarded people, alcoholics, career criminals, lower classes, and nonwhite races from breeding, thereby preventing the persistence of undesirable characteristics. Many European countries and US states had eugenic programs. It was only when the Nazis showed the end result of such reasoning with the Holocaust that we collectively realized, "Oh, hey, this is evil." Even so, it continued in some places.

So a large part of the objection to evolution was an objection to social Darwinism and its implications. But why did the good people of Tennessee, William Jennings Bryan, *The Fundamentals*, et al. equate biological evolution with social Darwinism? *For the same reason everyone else did.* They were seen as two sides of the same coin. "Survival of the fittest" was taken as a grand unified theory that applied to everything, not just biology. To be sure, there were people who didn't take it this way, but generally speaking, if you accepted biological evolution at that time you also accepted social Darwinism; and if you rejected social Darwinism you also rejected biological evolution.[38] Of course in retrospect we see that they should have made a distinction between the two, accepting biological evolution and rejecting social Darwinism, but it was an error made by both sides.

In fact, the textbook in question, *Civic Biology*, by George William Hunter, *explicitly taught social Darwinism and eugenics*. More than that, it taught them as *scientific discoveries*, mere extensions of the theory of evolution.

37. Livingstone, *Darwin's Forgotten Defenders*.
38. Sampson, *Six Modern Myths*, 62–70; Numbers, "Creationists," in Lindberg and Numbers, *God and Nature*, 394–98.

At the present time there exist upon the earth five races or varieties of man, each very different from the other in instincts, social customs, and, to an extent, in structure. These are the Ethiopian or negro type, originating in Africa; the Malay or brown race, from the islands of the Pacific; the American Indian; the Mongolian or yellow race, including the natives of China, Japan, and the Eskimos; and finally, the highest type of all, the Caucasians, represented by the civilized white inhabitants of Europe and America.

. . .

When people marry there are certain things that the individual as well as the race should demand. The most important of these is freedom from germ diseases which might be handed down to the offspring. Tuberculosis, syphilis, that dread disease which cripples and kills hundreds of thousands of innocent children, epilepsy, and feeble-mindedness are handicaps which it is not only unfair but criminal to hand down to posterity. The science of being well born is called *eugenics*. . . .

Studies have been made on a number of different families in this country, in which mental and moral defects were present in one or both of the original parents.

. . .

Parasitism and its Cost to Society.—Hundreds of families such as those described above exist to-day, spreading disease, immorality, and crime to all parts of this country. The cost to society of such families is very severe. Just as certain animals or plants become parasitic on other plants or animals, these families have become parasitic on society. They not only do harm to others by corrupting, stealing, or spreading disease, but they are actually protected and cared for by the state out of public money. Largely for them the poorhouse and the asylum exist. They take from society, but they give nothing in return. They are true parasites.

The Remedy.—If such people were lower animals, we would probably kill them off to prevent them from spreading. Humanity will not allow this, but we do have the remedy of separating the sexes in asylums or other places and in various ways preventing intermarriage and the possibilities of perpetuating such a low and degenerate race. Remedies of this sort have been tried successfully in Europe and are now meeting with success in this country.[39]

---

39. Hunter, *Civic Biology*, 196, 261, 263.

So it turns out that those closed-minded Christian fundamentalists were objecting to having social Darwinism and eugenics taught to their children as unassailable scientific facts. If we were to take this in isolation (which we can't), it looks like the ignorant, racist, southern hicks were taking a stand against racism and white supremacy.

Chapter 9

The God Killer

I. THE FOUR BECAUSES

BEFORE CONTINUING OUR DISCUSSION on evolution—which, I'm sure you'll recall, I accept—we have to return to ancient Greece and Aristotle.[1] When we ask for a cause of a particular phenomenon, we can give different answers depending on what we mean by "cause." Aristotle said there were four causes, or, as J. R. Lucas put it, "four becauses,"[2] four different types of explanation. These are not in competition with each other; indeed, a full explanation requires all four, according to Aristotle.[3]

First is the *efficient* cause. This is what produces the effect: it's what we mean by cause in the modern era and contemporary science, and this sometimes leads people to have trouble understanding the other three causes. Because our conception of causality is specifically efficient causality, they think when we refer to (e.g.) the formal cause we're referring to it *as* an efficient cause. Otherwise, why call it a cause at all? If that's a sticking point for you, substitute "explanation" for "cause": the four causes are four different aspects in explaining a phenomenon. Just remember that

---

1. If you really want to annoy a philosopher, pronounce Aristotle as if it rhymes with chipotle.
2. Lucas, *Freedom of the Will*, 34.
3. Aristotle, *Physics* 2.3, 2.7; *Metaphysics* 5.2.

the other three causes are not causes as we usually understand that term: they are not efficient causes, they are not what produces the effect.[4]

Second is the *material* cause. This is what the effect is made out of, it is the material the effect is composed of. A common example is a sculpture. The material cause of a sculpture is the marble block it is carved out of. The efficient cause would be the sculptor. At this point, it's fairly uncontroversial because most of us would recognize that you need something to cause the effect and some material for the effect to be made out of. So contemporary science affirms efficient and material causality, although it doesn't usually call the latter "causality," since that term is reserved for the efficient species.

Third is the *formal* cause. This is how the effect is related to its Platonic form.[5] The formal cause explains what it means for the effect to be what it is, to belong to the class it does. It's what makes the effect one object instead of a mere collection of atoms or molecules. The efficient cause of a table would be the person who built it, and the material cause would be the material it's composed of: wood or metal or whatever. But the formal cause is what makes it a *table* instead of just a mass of wood or metal. Basically, it's a table because it reflects the Platonic form of a table. This is not as clear as the other two, but it doesn't have anything to do with what I'm saying here, so you don't have to think about it anymore. You're welcome.

Fourth is the *final* cause (*telos*). This is why the effect exists, what it's *for*, what its purpose or function is.[6] So with a birthday cake, the efficient cause is the baker; the material causes are the ingredients (flour, sugar, etc.); the formal cause is whatever, I don't care; and the final cause is the celebration, the birthday party. This explains the effect because the efficient cause wouldn't use the material causes to produce the effect unless it had a reason to. If you tried to explain the making of a birthday cake without saying why the baker was making it, it wouldn't be a complete explanation.

So you can see why we would use these different concepts. Sometimes when we ask for an explanation of the birthday cake on the table, we're asking who made it. Sometimes we're asking what it was made out of (Is it chocolate? Lemon? Sauerkraut?). And sometimes we're asking why it was made. Again, these answers are not in competition with each other: giving the efficient cause doesn't deny the final cause. And according to

4. Falcon, "Aristotle on Causality."
5. See above, 88–89.
6. Aristotle, *Physics* 2.8.

Aristotle, a complete account of an effect would (or at least could) have all four. If you explain what flavor the cake is, you're not explaining who made it or why it was made.

In addition, Neoplatonists suggested two more kinds of causality: the *instrumental* cause, which explains how the efficient cause uses the material cause to produce the effect (for a cake that would be the bowl, spatula, pan, and oven); and the *paradigmatic* or *exemplary* cause, which is the plan or blueprint by which the efficient cause produces the effect (the recipe for the cake).[7] Others argue that these just collapse into the efficient and formal causes. But you get the idea.

A final point: there's an important reason why modern science focuses on the efficient and material causes to the exclusion of the others: only the efficient and material causes are *observable*.[8] Final causality is conceptual: we have to include *why* an effect took place in order to have a complete explanation, at least some of the time, even though we do not and cannot observe it as such. We can observe the birthday party, of course, but not *as* a final cause.

II. PRE-DARWINIAN EVOLUTION

Many people seem to think that the theory of evolution just sprang fully formed from Darwin's forehead. But not only were there forerunners to Darwin in his time and culture, there were forerunners throughout history, some more interesting than others.

Premodern ideas

Anaximander, a pre-Socratic philosopher from the sixth century BC, suggested that living creatures are formed from warm mud, and that human beings evolved from fish. However, it wasn't a gradual process "since

---

7. Helmig and Steel, "Proclus."

8. Well, Hume pointed out that we don't observe efficient causality either. We see one event followed by another event and automatically *infer* that the first caused the second. For example, we see a billiard ball moving across the table, coming into contact with another ball and stopping, and then we see the second ball moving away, effectively taking over the motion of the first ball. But, Hume argued, we don't actually see the first ball *causing* the second ball to move. All we see is the first ball moving and then the second ball moving. That's just a sequence of events; and while we immediately and correctly intuit that the first ball caused the second ball to move, technically that's not something we observe (Hume, *Enquiry Concerning Human Understanding* 4.1 [p. 13]).

Anaximander assumed that our ancestors did not have the fish as an ancestor but emerged from it fully formed."[9] Empedocles seems to have suggested something similar to natural selection to account for the extinction of plants and animals.[10] Aristotle thought everything could be placed on a *scala naturae* (natural ladder), what came to be called the great chain of being, with more complex things that exhibit clear functions (like living creatures) higher up than nonliving matter. However, this was more of a taxonomic point than an evolutionary one since he didn't suggest that the lower rungs eventually climbed up to the higher rungs.[11]

A better example is the theory of *rationes seminales* (seminal principles), which probably originated with the ancient Stoics. The idea is that God created living creatures in seed form, which then developed or unfolded according to their potentialities over time. This is blatantly evolutionary. It was very popular among ancient and medieval Christians, including Augustine, Bonaventure, Albertus Magnus, and Roger Bacon.[12] It fell by the wayside with Aquinas's theology. The fact that many early Christians accepted this form of biological evolution, including Augustine, who exhibited more influence on Christian thought than almost anyone else, is a significant point. Christian advocacy of *rationes seminales* goes back to the Roman Empire and the idea was commonly accepted for centuries. The point being that evolutionary concepts are not foreign to Christianity. However—not to spoil the plot—this idea uses final causality (in that it appeals to a potential that could be fulfilled), not just efficient and material causality.

In the ninth century, a Muslim author, al-Jāḥiẓ, wrote the *Book of the Animals* (*Kitab al-Hayawan*). In it, he seems to foreshadow Darwinian evolution, at least according to some. Mehmet Bayrakdar writes that Jāḥiẓ "described three mechanisms. These are Struggle for Existence, Transformation of species into each other, and Environmental Factors."[13] This is true and valuable. However, the human race had been agrarian since time immemorial, and anyone living in an agrarian society knows perfectly well about the struggle for existence and how environmental factors play into it. (Only the strongest of the cattle will survive during

---

9. Ring, *Beginning with the Pre-Socratics*, 29; cf. Copleston, *History of Philosophy*, 1:25.
10. Ring, *Beginning with the Pre-Socratics*, 143.
11. See, e.g., Aristotle, *History of Animals*.
12. Copleston, *History of Philosophy*, 2:75–78, 274–78, 298, 448.
13. Bayrakdar, "Al-Jahiz and the Rise of Biological Evolution," 151.

a drought when the fields in which they feed provide little food, for example.) Adding the transformation of species into the mix, though, points the way towards biological evolution. However, Jāḥiẓ's claims on this do not indicate anything other than small changes within set limits. This, also, was already believed from time immemorial by Muslims, Christians, and Jews, who claimed that all the varieties of human beings are descended from Adam and Eve. However, they may have been held unconsciously. Jāḥiẓ brought these ideas to the forefront and developed them into a nascent and limited form of natural selection, but there's no indication that he tied them together into an evolutionary theory.

This is not to say that Jāḥiẓ's contributions are not significant, just that they are being exaggerated.[14] He seems to have been more of a taxonomist, like Aristotle, than an evolutionist. The best quote from Jāḥiẓ foreshadowing Darwin that I've found is from a news article,[15] but I can't find anything remotely like it in the English translation; and academic references of Jāḥiẓ presaging Darwin don't even seem to be quoting the same passage.[16]

Modern ideas and Lamarckism

In the modern era, evolutionary ideas were suggested more often. Remember, there was an innate sense of developmentalism throughout the eighteenth and nineteenth centuries, the idea that we were orchestrating our own return to Eden, as it were, and that the advancement of science was the tool by which we were going to immanentize the eschaton. So it was tempting to see this development, this perfecting process, everywhere in the world on all levels and in all contexts, including biology. In the eighteenth century, evolutionary-ish ideas were proposed by Benoît de Maillet, Pierre Louis Maupertuis, Denis Diderot, George-Louis Leclerc, James Burnett, Erasmus Darwin (Charles's grandfather), and Montesquieu, but none of them had a method or mechanism to make it work.[17]

Then along came Jean Baptiste Lamarck. He proposed a method by which living creatures, in their struggle for survival, would produce new

14. Egerton, "History of the Ecological Sciences," 143.
15. Al-Khalili, "Science: Islam's Forgotten Geniuses."
16. Pellat, *Life and Works of Jāḥiẓ*, 130–85; Zirkle, "Natural Selection Before the 'Origin of Species,'" 84–85.
17. I don't want to provide references for all these people. Just google them.

morphological traits, which they would then pass down to their progeny. Since these traits were produced to allow the organism to survive better, they gave the organism's offspring an advantage.[18] The giraffe's neck is a common example. Once upon a time, giraffes had much shorter necks. But then they became so numerous that they ate all the leaves within range of those necks and started starving. One particular proto-giraffe stretched its neck as hard as it could and lengthened it enough so that it had access to leaves the others couldn't reach. This giraffe survived and passed on its slightly longer neck to its children. Pretty soon, proto-giraffes with that length neck were all that was left, the shorter-necked ones having died off from being unable to get enough veggies in their diet. But then the same scenario arises, and one more proto-giraffe stretches his neck a little further, and it happens all over again. And again. Eventually you get the freak show we all know and love as the giraffe. You can run through similar scenarios to explain the particular morphological traits of all living creatures, from single-cellular organisms to penguins, kangaroos, and the majestic moose.[19]

Lamarckism was very popular. It fulfilled the earlier proposals of Maillet, Diderot, and others but *explained* how it happened. We now know that the explanation doesn't work because a parent can't pass on acquired traits like this to its offspring, but they didn't know that then. In fact, at the time there were widespread accounts of it really happening: a father loses his arm in the factory, then his son is born missing the same arm; a woman suffers from some kind of burn, then her daughter is born with a birthmark in the exact same place and shape as her mother's scar; that kind of thing.

So let's go back to the four causes. The other evolutionary proposals did not write God out of the picture, and what kept him in was final causality. Why does this animal have those properties? Because God set things up so that it would have traits that matched its environment, so that those traits had a particular function or purpose or end goal. Lamarckism is different. It still has final causality, but it displaces it from *God* to the *organism*. Why does this animal have those properties? Because the organism changed itself so that its traits matched its environment. Granted, it wasn't doing this *consciously*, it was just trying to not die (a good goal to aspire to), but that's still final causality. Some people

---

18. Packard, *Lamarck, the Founder of Evolution*.
19. A Møøse once bit my sister.

objected, mostly on scientific grounds, such as August Weismann and Georges Cuvier, but others accepted and defended Lamarckism, including Étienne Geoffroy Saint-Hilaire and Robert Edmond Grant. Theoretically, Lamarckian evolution could explain how the diversity of species came about from a single original organism without appealing to God, but that didn't amount to a reason to *exclude* God. It wasn't an argument *against* God's existence, although it did challenge the need to appeal to God in a certain arena.

Any evolutionary explanation is going to need an account of how differences (mutations) are produced and an account of how these differences were selected. The mutations can be rational (that is, goal directed), like the case with the proto-giraffe, or natural (*not* goal directed). Similarly, the selection process can be rational or natural. Since both steps can have either rational or natural explanations, this gives us four possibilities.[20] If either half is rational, then it includes final causality, since I'm defining rational as goal directed.

1. Rational mutations and rational selection
2. Natural mutations and rational selection
3. Rational mutations and natural selection
4. Natural mutations and natural selection

Nicholas Rescher suggests that Teilhard de Chardin is a representative of 1 and Henri Bergson is a representative of 2, because of Bergson's proposal of an élan vital (vital impetus or force) that guides the selection process.[21] Lamarckism is 3: the mutations are produced intentionally—that is, the organism's struggle to survive produces the morphological change—but the selection process doesn't have to be. This is because the mutations are produced *in order to* (final cause) make the organism more adapted to its environment. To be sure, if we observed those Lamarckian proto-giraffes we wouldn't see the final cause, because final causality is not observable, it is conceptual. But Lamarckian evolution requires it.

---

20. Rescher, *Useful Inheritance*, 7–8.
21. Bergson, *Creative Evolution*.

III. DARWINISM AND FINAL CAUSALITY

Charles Darwin's theory of evolution has other predecessors than those already mentioned. His grandfather, Erasmus Darwin, was very influential on him, as was Robert Chalmers's 1844 book *Vestiges of the Natural History of Creation*, which argued that there is a natural trend to produce order out of chaos and this applies to all facets of the universe. But Darwin improved upon them and on Lamarck by coming up with a mechanism (not just a method) to explain the diversity of species that did not involve final causality *at all*; that's option 4 above.

We do not need to appeal to God to explain how mutations are produced by creating the organisms so that they fit into their environment, nor to the organisms and their struggle for survival. They are produced randomly. "Randomly" in this context does not mean uncaused but that the causes of the mutations are disconnected from the selection process: they are not produced *in order to* make the organisms more capable of being selected (that is, surviving).[22] As for the selection, this is a natural process too, not intentional, not brought about by Bergson's élan vital or anything else. So Darwinian evolution involves natural mutation and natural selection. This removed final causality entirely—or, rather, it explained the origin of species without any need to *appeal* to final causality.

Traditionally, it was thought that the production of complex forms of life could not plausibly be explained by natural processes, so it required referring to God and his intentions. Lamarckism removed the need to appeal to God to get final causality, and Darwinism removed the need to appeal to final causality at all. This is why evolution is the prime example people use when they argue that science is slowly weeding out religion. It shows, allegedly, that despite the widespread belief that certain phenomena could be plausibly explained only by supernatural processes, they could, in fact, be explained by natural processes. This is held up as a *pattern*, and evolution is the exemplar of it.

Of course it's true that science corrects things this way sometimes; whenever you find a new way to explain things, a lot of the old explanations are going to be corrected. But science also corrects things the other way by showing that matters we thought were easily explicable by natural processes are resistant to such explanations. This isn't usually appreciated as much, but I'd like to suggest the origin of life (abiogenesis) as the exemplar here. Prior to the twentieth century, people thought simple forms

---

22. Sober, "Evolution Without Naturalism," 3:191–96.

of life commonly came into being from nonliving matter by natural processes, called spontaneous generation.[23] But while they thought life could arise through natural processes, they saw no pathway from those simple forms of life to the complex ones, so the latter had to be created directly by supernatural dictate.

So living creatures coming into existence naturally from nonliving material was considered a simple, natural process, one that was commonly observed. But for decades now explaining the origin of life naturalistically has been shown to be more and more intractable. Speaking as a nonscientist, I have always been partial to Cairns-Smith's clay hypothesis,[24] but, like all the other theories, it has shown itself to be woefully insufficient so far.

I don't want to press this point too hard because I expect there *is* a naturalistic origin of life. But the larger point I'm making is that the idea that there is this pattern of science explaining things naturally that were thought to be too difficult to be explained that way is an exaggeration. It sometimes goes the opposite way. If there is a pattern here, it's nowhere near as one-sided as is claimed. Why should we expect anything different? When science corrects misconceptions, those corrections can go in any direction. To insist it nearly always goes the same way, or at least that it *should*, is to use science as an ideology instead of a tool for discovering truth. Anyone who has studied the history of artificial intelligence knows that many things we thought would be nigh-impossible were relatively simple, and many things we thought would be simple turned out to be nigh-impossible.

IV. REMOVING GOD

Does Darwinism really remove God from the evolutionary process though? We don't have to *appeal* to God anymore, but that was true of Lamarckism as well. With Darwin, however, we don't even need to appeal to teleology, and that's a pretty significant development, even though final causality is, again, not observable but a conceptual element of (at least some) explanations.

---

23. McCartney, "Spontaneous Generation and Kindred Notions in Antiquity"; Farley, *Spontaneous Generation Controversy from Descartes to Oparin*.

24. Cairns-Smith, *Seven Clues to the Origin of Life*, 80–86.

But that still doesn't answer the question: Does Darwinism remove God from the evolutionary process? My answer is no, and I'll make this point a few ways.

God and the laws of nature

First, the traditional claim of Judaism, Christianity, and Islam has always been that God usually acts *through* the laws of nature that he set up. In this case we'd say that these laws are the tools or the instrumental cause of the effects. This is not some attempt to reintroduce God into an area from which science has banished him; this has *always* been the theist's claim.

Example: in the Bible God parts the Red Sea so the Israelites can pass through before being killed by the Egyptians (passing through *after* being killed by the Egyptians wasn't an option apparently). Whenever this event is portrayed on film, it's always made out to be some dramatic supernatural event: Moses stomps down his staff and the waters flee away or something. But the actual text doesn't suggest this. For one thing, most scholars think the Red Sea in question (*yam suph*) was the Sea of Reeds, a swamp or mangrove somewhere in the area where the Suez Canal is today. There's a book by Maryse Condé titled *Crossing the Mangrove*, the theme of which is that you *can't* cross a mangrove. It's impossible. Condé uses this as an analogy of translating the spoken word to the written word, and then from one language to another. Each translation process kills the story and creates a new and different story in its place.[25]

So there was a swamp that was uncrossable, but still there had to be a miracle to move the water so that the Israelites could cross through anyway, right? Well, the text simply says that Moses held up his staff and a strong east wind started blowing. After about twelve hours, enough water had been blown away that the swamp became passable. There have been studies saying something like this could very well take place in that area,[26] although since the swamp is no more, it's speculative. And—assuming this event happened in the first place, which I'm not insisting on—this *could* very well have been a miraculous event. But it could also have been a natural event caused by God, since that is how God usually

---

25. The translations in the story go from spoken Creole to written Creole, then from Creole to French. I know this book by its translation from French into English, so that's one more killing.

26. Drews and Han, "Dynamics of Wind Setdown at Suez and the Eastern Nile Delta."

acts, and that seems to be what the text is describing. In this latter case, we would be able to explain it without any appeal to God, just to the natural processes in play.

Once at a philosophy conference, I was talking to a guy about this: I said, if the scientific evidence became clearer and was able to demonstrate that such an event could happen by natural processes, would this conflict with the Bible's claim that *God did it by natural processes*? He said yes. Since the natural processes explain the alleged phenomenon, any appeal to God becomes gratuitous or ad hoc.[27] I asked that if the scientific evidence confirms what the Bible says, how would this be evidence *against* the Bible? We both stuck to our guns, but I'm just going to say that he was wrong, and I was right. If it were gratuitous or ad hoc to say God was involved in a natural process simply because it's a natural process, then this would mean that scientific confirmations of biblical claims would constitute evidence *against* those biblical claims which is . . . well, let's just say strongly counterintuitive. And since it is the traditional view of God that he usually acts through the natural processes he set up in the first place, there is nothing inappropriate or contrived in saying that explaining phenomena via natural processes does not exclude God from them. This could be *wrong* of course—God may not exist or the Bible may be wrong about how he usually works—but those are different questions.

So if we want to read natural processes, including evolution, as excluding God, we have to interpret them atheistically. Evolution + atheism entails that God was not involved because it entails that God does not exist. But that's not too surprising. Newtonian physics + atheism entails that God does not exist. Seventeenth-century French shipbuilding + atheism entails that God does not exist. *Anything* + atheism entails that God does not exist. So in order to make evolution incompatible with Christianity, you have to supplement it with atheism; and then it's obviously the supplement that creates the problem, not what it's supplementing.

What is Darwinism excluding?

Second, the lack of an appeal to final causality in Darwinism is more restricted than is sometimes presented. The idea is that there is no physical mechanism, either in the organism or its environment, that foresees just which mutations would be beneficial and then causes those mutations

---

27. See below, 205–8.

to take place.[28] This is unlike the teleological forms of evolution where the two halves, the source of mutations and the selection process, *communicate* with each other in a sense so that the mutations are meant to make the organism fit its environment better and so increases its survivability. But this is not the case for Darwinism, in which there is no physical mechanism uniting those two halves. But this is an important constraint, that it restricts itself to physical mechanisms. It restricts itself this way because any nonphysical causes in play would not be observable and any cause that acted non-mechanistically would not allow the inferential inductions from particular to general on which science depends. Such causes would allow only observations that could not be generalized into an explanation that applies to similar scenarios. And God is neither physical nor a mechanism. Thus the claim that Darwinism posits no connection between the source of mutations and the selection processes does not apply to God. And we can't change the definitions so that they do apply to God for the reasons already given: God is neither physical nor a mechanism, and so is not directly accessible to scientific analysis. As John Henry Newman, one of the most celebrated theologians of the nineteenth century, put it, "I do not [see] that 'the accidental evolution of organic beings' is inconsistent with divine design.—It is accidental to *us*, not to *God*."[29]

The response to this is that Darwinian evolution still renders God *gratuitous* in that he is not a condition required to make sense of the phenomena. To say that God is a part of the process violates Ockham's Razor, which says simpler explanations are generally preferable to more complex ones. More specifically, the razor says that, all other things being equal (which they never are), we should prefer an explanatory hypothesis that involves fewer entities to one that involves more. So "simpler explanation" isn't a qualitative term, it's a quantitative one. Generally speaking, the numerically simpler explanation is preferable.

I think this objection is true in a sense, since we don't need to posit God and his intentions (teleology) to account for the development of the species. But there are counterresponses that take the teeth out of it. First, Ockham's Razor is the claim that we should not multiply entities when formulating an explanatory hypothesis. But those who think God is nevertheless involved in the evolutionary process in the aftermath of

---

28. Sober, "Evolution Without Naturalism," 3:191–96.
29. Newman, *Letters and Diaries*, 77–78; emphasis in original.

Darwin are not usually offering this as part of an explanatory hypothesis. They're not suggesting God's action must be posited to explain the phenomena; they have different reasons to believe in God and that he is involved in everything. We don't need to appeal to him to explain how a carburetor works either, but this doesn't make carburetors evidence of his nonexistence.

This can be shown via analogy. Take a player piano (i.e., a self-playing piano): we can explain its workings according to its internal processes without appealing to anything else outside it. Those processes fully explain what the piano is playing and how it's doing it. The internal processes of the piano parallel the laws of nature, and the melody it plays parallels living creatures. But surely it's not gratuitous to appeal to the person who built the player piano, as well as the person who invented it, and the composer who wrote the melody it's playing. This is because we want to know *why* the piano's inner workings are what they are so that they produce the melody; simply showing *how* it works leaves a lot of questions unanswered. In a similar way, we want to know why the laws of nature are what they are so that they produce complex living creatures. This doesn't detract from wanting to know how they do so, it's just a further question. So evolution pushes the problem one step back. It explains how living creatures evolved *given* the laws of nature being the way they are; but that's *not* given because the laws of nature could be different. We need an explanation of why they are the way they are.

Of course, we can ask the same question of God: What explains his (alleged) existence and why he has the particular properties he has? The number of people who think that this question has never occurred to theists is staggering. This is an example of what I started this book with: thinking our knee-jerk reactions are cleverer than the lifelong reflections of the most intelligent people who have ever lived. Anyway, the answer is that a supreme being would be *metaphysically necessary*, otherwise it would not be *supreme*. So the reason why God exists instead of not, why he has the properties he has instead of others, is because he's metaphysically necessary. He exists and has those properties in all possible worlds; it's impossible for him to not exist or to have different properties.[30] Note that this is not an argument *for* the existence of God, it's just saying that *if* God exists, he exists and has certain properties in all possible worlds. The flip side is that if God doesn't exist, he doesn't exist in *any* possible world.

---

30. Plantinga, *Nature of Necessity*; Leftow, *God and Necessity*; Pruss and Rasmussen, *Necessary Existence*.

God is either metaphysically impossible or metaphysically necessary: he either can't exist or he can't not exist. This is relatively uncontroversial in philosophy (except insofar as *everything* is controversial in philosophy).[31] And the answer to the objection "Why does God exist instead of not?" is that God would be metaphysically necessary. In fact, this is a better explanation than appealing to something else as an explanation of why something exists, because in the latter case we'd have to ask the same question of the second something that we asked of the first something. But if something exists necessarily, its existence is fully explained.[32]

Biology and teleology

Third, one aspect of early modern science was the recognition that we don't need to appeal to final causality in the realms of physics and chemistry (although this was disputed). Darwin showed we don't need to appeal to final causality in the evolutionary process and the origin of species—which had hitherto been the apotheosis of biology not being able to remove final causality. If you could remove it, as Darwin seemed to, then all the merely problematic cases in biology would fall in line and that right quick. However, as noted above, science goes in all directions, not just the one you want: the inexplicable becomes easily explicable, and the easily explicable becomes inexplicable.

My point is that *final causality is still all over the place in biology*, as well as all the life sciences and social sciences.[33] For example, when we ask what the heart is, we can answer by saying what it *does*. So we can say the heart constricts methodically. But then we would want to know why it does this: Why would an evolutionary process select a constricting muscle? Because by constricting it circulates blood throughout the body. Again, we'd want to know why it does this, and the answer is to circulate *oxygen* throughout the body. We can keep asking, and eventually we'd end up with saying the heart does what it does to promote the health and survival of the organism. But this is to say what the heart is *supposed* to do. It's how we tell the difference between a healthy heart and a diseased one; between one that's functioning properly and one that's

---

31. Craig, *Reasonable Faith*, 185.
32. Swinburne, *Existence of God*, ch. 4.
33. Nissen, *Teleological Language in the Life Sciences*; Plantinga, *Warrant and Proper Function*, 4–6.

malfunctioning. According to Ernst Mayr and George C. Williams, virtually every advance in physiology has come from asking what a given organ or structure is supposed to do.[34]

This isn't reducible to saying what things usually do. You've probably seen videos of those adorable baby sea turtles hatching on the beach and making their way to the ocean. Most don't make it; they're picked off by predators. They don't usually show that on the videos. Most of those that *do* make it get picked off by predators in the ocean. The percentage of sea turtles that ultimately survive to adulthood is tiny, much less than 1 percent. If we equated functions with what normally happens, we'd have to say that the function of sea turtles is to die violently immediately after hatching, and those lucky few that make it to adulthood are malfunctioning. This is obviously not the right answer. Nature is set up to overprovide in this way. The function of sperm cells is to fertilize an ovum, but how many actually do?

Biological functions, what certain organs, organisms, or structures are supposed to do, are present throughout biology,[35] and Darwinian evolution doesn't remove them. We can certainly hope that further discoveries will do so, but that's all it is: a hope. To expect that since one area where we were able to remove the need to appeal to teleology is indicative of a general trend is to take science as an ideology. It's just as likely that we will find areas where we have to *add* it. There are other options: Daniel Dennett has tried to make evolution into an algorithm that applies to all biological subfields as well as beyond biology,[36] but it's a very contentious point since it doesn't really work; it's still merely a hope. Others have tried to make sense of biological functions within a naturalistic framework, but none of these seem to work either.[37]

Perhaps we could say that this teleology is only apparent, not actual, so that we can take functionalism as fictionalism.[38] Other sciences frequently use idealized conditions or entities that do not exist to explain the phenomena that do (remember how Copernicanism was thought to explain the appearances without being physically true). We can *try* to do

---

34. Mayr, "How to Carry Out the Adaptationist Program?," 328; Williams, *Plan and Purpose in Nature*, 22.

35. Melander, *Analyzing Functions*.

36. Dennett, *Darwin's Dangerous Idea*, 48–60.

37. Plantinga, *Warrant and Proper Function*, 194–215; Plantinga and Tooley, *Knowledge of God*, 20–30.

38. See above, 48–49.

this with biological functions, but the consequences are severe. Saying that ideal gases aren't real doesn't challenge much. If we say that biological functions aren't real, we are saying the difference between a healthy heart and a diseased one is not a real distinction; the difference between a healthy organism and a dead one is just an illusion. That is very difficult to take seriously, and even if we could, it would be on Hume's level where we can believe it only in the quiet of our studies and not when we are out and about in our daily lives.[39] Worse than that: if our brains don't have the function of forming true beliefs in part, if that's not what they're *supposed* to do, this produces a reason for us to distrust the contents of our own minds—and obviously, this would include the beliefs that biological functions aren't real, that evolution is true, that human beings don't usually emerge out of cocoons, etc.[40]

V. DAMAGE ASSESSMENT

So what damage does evolution do to religion? The only correction it offers is that it makes a set of phenomena the product of natural acts of God rather than supernatural acts of God. Why not natural acts *without* God? That would be evolution + atheism, not evolution *simpliciter*. That's certainly possible though: Darwinism encompasses evolution + atheism as well as evolution + theism, evolution + neoclassical economics, and whatever else you want to add to it. But obviously it's the "+ atheism" part that produces the problem we're looking at, not the "evolution" part.

I think there are three issues where evolution says something about Christian or theistic concerns. I'll put them in order of "least damage done" to "most damage," where the deciding criteria are basically what I find most interesting. Ironically, both sides tend to accept these issues, they just disagree which side should win. So by diminishing the tensions between them, I should be infuriating *everybody*. That makes me feel all warm and fuzzy.

---

39. Hume, *Treatise of Human Nature* 1.4.7.
40. Plantinga, *Where the Conflict Really Lies*, 307–50; Plantinga and Tooley, *Knowledge of God*, 30–51; Jim Slagle, *Evolutionary Argument Against Naturalism*.

Evolution and the Watchmaker Argument

The issue most people focus on is that evolution repudiates one of the teleological arguments for the existence of God, namely the Watchmaker Argument. This argument compares the complexity of living creatures to the complexity of a watch, and just as we would infer that an intelligent agent built the watch, so we should conclude that an intelligent agent created living creatures.[41] Evolution provides an explanation for the complexity of living creatures without needing to appeal to an intelligent agent. But refuting an argument is not the same thing as showing that its conclusion is false. All it does is show that the argument *fails to demonstrate* the conclusion. The conclusion could still be true, and if there are other arguments for it, we can still have good reasons for accepting it. Evolution doesn't constitute an argument *against* God's existence, just a rebuttal of an argument *for* God's existence. In fact, it rebuts only one application of the Watchmaker Argument: it doesn't rebut its application to the origin of life. Plus, there are other types of teleological arguments out there, like those based on the anthropic coincidences.[42] And there are other families of arguments for the existence of God, like cosmological arguments, moral arguments, ontological arguments, and others that evolution doesn't touch. So evolution rebuts one application of one type of one family of theistic arguments. I don't want to belittle this, but I have to say that if that's the worst evolution does, I'm thoroughly underwhelmed.

Evolution and the Bible

I *say* that's the issue most people focus on, but for many the real issue between evolution and Judeo-Christianity is that it conflicts with the particular story of creation in the Bible. The first such claim is that evolution is incompatible with God taking six days to create the universe and then resting on the seventh day, which wouldn't leave enough time for evolution to work. This assumes that the six days of creation and God's Sabbath day of rest should be understood as calendar days (or solar days or human days or "normal" days). But this is at least contestable, and at any rate, it wasn't a common interpretation in Darwin's own day. The most common interpretations then were the gap and day-age interpretations

---

41. Paley, *Natural Theology*.

42. Collins, "Teleological Argument." For a couple of examples, see above, 60–61, 63–64.

as opposed to the calendar-day interpretation.[43] Another common take is the literary framework interpretation, which argues that, since the events of days one, two, and three seem to correlate to those of days four, five, and six respectively, Gen 1 is not describing a sequence of events at all. These four interpretations—the calendar-day, gap, day-age, and literary framework—all predate scientific evidence for the age of the universe, and three of them date back to the ancient church, so they were originally presented and accepted for purely biblical reasons, not out of an attempt to reconcile the Bible with science. Obviously, there is *much* more that could be said here as it's one of the most visible areas of tension between science and religion, but it would require its own book.[44]

In addition to the days of creation, there are also plenty of charges that the evolutionary story about human origins supposedly contradicts the biblical story of Adam and Eve's creation in the garden of Eden. The first thing to say in response to this is that there have been people who have taken the early chapters of Genesis as nonhistorical parables or some such since the early church. In a text probably written in the AD 220s, Origen writes,

> Now who is there, pray, possessed of understanding, that will regard the statement as appropriate, that the first day, and the second, and the third, in which also both evening and morning are mentioned, existed without sun, and moon, and stars—the first day even without a sky? And who is found so ignorant as to suppose that God, as if He had been a husbandman, planted trees in paradise, in Eden towards the east, and a tree of life in it, i.e., a visible and palpable tree of wood, so that anyone eating of it with bodily teeth should obtain life, and eating again of another tree, should come to the knowledge of good and evil? No one, I think, can doubt that the statement that God walked in the afternoon in paradise, and that Adam lay hid under a tree, is related figuratively in Scripture, that some mystical meaning may be indicated by it. The departure of Cain from the presence of the Lord will manifestly cause a careful reader to inquire what is the presence of God, and how anyone can go out from it.[45]

---

43. Numbers, *Creationists*, 7–8.

44. For a debate between advocates of the calendar-day, day-age, and framework interpretations, see Hagopian, *Genesis Debate*.

45. Origen, *De Principiis* 4.1.16.

Now, as far as I can tell, most Christians *did* take these stories historically through the ages, despite what Origen writes here. My point is just that it wasn't a settled issue. Even ignoring this, what exactly is the contradiction between the evolutionary account and the biblical account? The Bible says God made the first human beings out of the dust of the earth; evolution provides some of the intermediate steps this took.

One specific way they could be made inconsistent is that most biologists say we are descended from a group of hominids, not from two individuals. But there are plenty of ways to accommodate such an idea into the biblical text without denying its historicity. Bear in mind, I'm not *claiming* anything about its historicity; I'm just saying that *if* we have to take the biblical account as historical, which we don't, it's not strictly incompatible with the evolutionary story. Granted, there isn't much in either account that suggests the other, but so what? That only means that any attempt to make a story that includes both of them would be completely contrived to those who accept only one. This, in turn, means that we can't use an attempt to reconcile the two accounts as evidence for accepting both. But that's pretty weak tea.

Evolution and evil

Arguments from evil claim that the existence or occurrence or degree of evil in the world constitutes a reason to not believe in God, and they make up the large majority of arguments (as well as knee-jerk reactions) against the existence of God.[46] Attempts to reconcile the concept of an omnipotent and morally perfect God with evil are called theodicies.[47] Some people use evolution to bolster up arguments from evil, since it shows that there has been much, *much* more evil and suffering in the world than people used to think. The only certainties in life are two particularly nasty evils: death and taxes. But does this show that the various theodicies don't work? Alvin Plantinga points out that such an objection would work just as well going into the future as the past.

> Suppose we learn that our world, with all its problems, heartaches, and cruelty, will endure for millions of years before the

---

46. The literature on the problem of evil is endless. I recommend Tooley, "Does God Exist?" in Plantinga and Tooley, *Knowledge of God*, and Marilyn McCord Adams, *Horrendous Evils and the Goodness of God*, to get both sides.

47. Theodicies often follow theiliads.

advent of the New Heaven and the New Earth; that wouldn't have much bearing, so one thinks, on the viability or satisfactoriness of this response to evil. (The New Heaven and the New Earth, after all, will exist for a vastly longer period than our current sad and troubled old world.) But the same goes, I should think, for our learning that our world with all the ills it is heir to, has gone on for much longer than originally thought. Current science shows that suffering, both human and animal, has gone on much longer than previously thought; but it doesn't thereby diminish the value of Christian responses to the problem of evil and in this way doesn't exacerbate that problem much, if at all.[48]

Well, it doesn't if you're just looking at the amount or extent of evil. But the challenge of evolution is regarding the *origin* of evil. Perhaps a perfectly good God could allow evil, and even use it to bring about good, but how could he be the instigator of evil and still be perfectly good? Judeo-Christianity supposedly explains the origin of evil as a consequence of the first human beings using their free will (which is a good thing) to reject God. This is how evil and death entered the world. Evolution rebuts this explanation by showing that evil and death predate humanity entering the scene by billions of years.

Now, I say this is how Judeo-Christianity *supposedly* explains the origin of evil because it's another misunderstanding. Here I'll make two biblical points. First, according to the story of Adam and Eve, evil already existed and had access to the world, including paradise, before they had sinned. That's precisely how they were tempted to sin in the first place, by Satan taking on the form of a snake and tricking them. So the idea that the Bible says that there was no evil until the first human beings sinned is simply incorrect. The reason so many people on both sides think otherwise is because they think the Genesis story is trying to show the origin of human death, thus the origin of animal death (so no evolution), and thus the origin of evil. But applying this to anything other than human death goes beyond what the text says—and it's not obvious whether it's referring to spiritual death (separation from God), physical death, or both. Moreover, Jesus said that Satan "was a murderer from the beginning" (John 8:44), which hearkens back to the creation of the universe ("In the beginning . . ." [Gen 1:1]), not just the part after human beings fell. Christian tradition lays the origin of evil upon Satan's fall from grace, not humanity's, but that's not explicitly stated in the Bible.

---

48. Dennett and Plantinga, *Science and Religion*, 11.

So what can we conclude from this? That Satan introduced death into the world and God allowed it and even used it to accomplish his good purposes? Maybe. From a Christian or Islamic perspective, there's nothing intrinsically implausible about that, but if you come at it without having beliefs about God (much less Satan) it's going to sound silly. If you're trying to convince someone that Christianity is true, I wouldn't lead with it. The more you parse it out, the less plausible it sounds (to me, at least), but I think it's enough to say that the Bible gives us a strong reason to think that evil and death were in the world "from the beginning" and that God is not the one responsible for it.

There are other ways to show that evolution does not call God's goodness into question, that it doesn't make him the author of evil or whatever, but that's enough for now. Fun subject, eh?

Chapter 10

Putting People in Their Place

I. THE MYTH OF MORTIFICATION

Another urban legend about science and religion is one we've already had a partial introduction to. This is the myth that science is going against religion by showing that we humans are not as important as we thought we were. We used to think we were valuable because religion says we are, but science has come along as the great iconoclast, showing us that we are completely unimportant. Science is progressively mortifying vain human pretensions to value and significance. Remember Sagan's reference to "the series of Great Demotions, downlifting experiences, demonstrations of our apparent insignificance, wounds that science has, in its search for Galileo's facts, delivered to human pride."[1]

The first case of this is the Myth of Dethronement, which equates geocentrism with egocentrism. We discovered that the earth is not at the center of the universe, then later discoveries showed that the sun is the center of only our solar system, not the universe, that the galaxy is not the center, and ultimately that *nothing* is at the center. Asking what's at the center of the universe is kind of like asking what point on the surface of a sphere is at its center. The problem with this decentralization movement, as discussed in chapter 3, is that the premoderns thought that the center of the universe was the least prestigious, least honorable, and most degraded place to be. This is why they said hell was at the center of the

---

1. Sagan, *Pale Blue Dot*, 26.

earth and thus the universe. If we swap out infernocentrism for geocentrism it becomes much easier to see that being at the literal center was not equated as being at the most important place.

Even on the earth's surface Christians considered themselves at the worst place. The T-O maps that were used to represent the known world in the Middle Ages were drawn with East up and the garden of Eden at the top.

> Thus the Christian map-reader orients himself (hence "the orient") towards the past paradise, emphasizing his current sinful nature. Many later maps include an image of Jesus at the top outside the landmasses and waters, thus also orienting the penitent towards potential future salvation. Europe, in the Southeast near the bottom of the map, is therefore far from the godhead and godliness, as fallen and far from redemption as possible.[2]

(I've lived in Europe, it's not that bad.)

Being immobile at the universe's center was also problematic because motion was a good thing. They believed that the stars and planets moved because they were good. The reason the earth was motionless at the center of the universe (the only place where it *could* be motionless) is that it wasn't good enough to move. Moreover, *light* was a good thing too: they believed that all the heavenly objects shone, either producing their own light or reflecting it. But not the earth, and for the same reason it didn't move: it wasn't good enough to shine.[3] The earth was stuck in the negative state of immobility, at the very worst location in the entire universe, and didn't even have enough value to reflect light, much less produce its own.

The second case is the Big Fish in a Small Pond Myth. People could have thought they were important when they thought the universe was tiny and that they (or the earth, rather) were the largest thing in it. But, per chapter 4, they already knew that the universe was larger than they could fathom and that the earth was, for all practical purposes, an infinitely small point of zero volume within it. To be sure, we've discovered that the universe is even *more* unimaginably large, by many orders of magnitude. But that doesn't mean they were starting from a conception of the universe as small. It's only small by comparison, not objectively.

---

2. Goldie, *Idea of the Antipodes*, 42.
3. C. S. Lewis, *Miracles*, 57.

They thought the earth was an infinitesimal speck in a cosmos vast beyond our abilities to understand or imagine.

So this Myth of Mortification is a nonstarter, since it says that science corrects a misconception that people didn't have. To be sure, they *did* think we were at the center of the universe and that the universe was orders of magnitude smaller than we have discovered it to be. But they did not think they were important or valuable by virtue of their relative size to the universe's or their location within it. More than this, though, what is the *connection* between size and location to value or significance? I suspect this idea is based on a utilitarian sense of value. In this case, our cosmic significance is tied up with how big we are in relation to the universe, since this says how much of an impact we can have on the universe. The bigger we are, the more centrally located we are, the more things we can influence; and the more things we can influence, the more important we are. Even on those terms, though, I don't see the connection: superstrings, if they exist, seem to be important. Life seems to be important. Rationality seems to be important, since by it the universe has become aware of itself[4]—and that holds regardless of how much space it takes up or its cosmographical location.

II. VALUE

However, for this to be a case of science conflicting with religion we have to ask what religion means by value. Specifically, what is value in the Judeo-Christian tradition? This is a deep question, but fortunately I've had to dumb it down for myself to understand it, so now I can share this dumbing down with you.

There are two sources of value within Judaism and Christianity. First is being loved by God. God is the ground of existence, the ground of morality, the ground of rationality, and the ground of all value. To be valued by the ground of value is to have objective value. Moreover, this love is not limited by having many objects, since the source has no limit. If you have kids, you understand: the more kids you have, the more you have to distribute your attention between them, not to mention patience, because you have only a finite amount of both. But this doesn't apply to love. It's not as if you love your first kid less because you have only five hundred love units

---

4. Carl Sagan said something like this in his *Cosmos* series and it's always resonated with me.

to express, and to give the new kid the right amount, you have to reduce the first kid's allowance. With God, love, value, and attention are unlimited. God's paying attention to others does not lessen the amount of attention he pays to you, and this goes just as much for his love.

OK, so according to Judaism and Christianity God loves us. This is important, but it doesn't get us very far, since God loves everything, according to these traditions. He may love some things more than others, and perhaps we can say the more God loves something, the more value it has. But setting up a ladder of value based on how much we think God loves things is hazardous—not least because people would inevitably use it as a reason to *disvalue* those things on the rungs below them.

If we create such a ladder and base it on what God created more of (in the physical universe, at least), we'd come up with

1. Empty space
2. Matter
3. Life
4. Rationality

There's more empty space than matter (occupied space), there's more nonliving matter than living matter, and there's more nonrational life than rational life. But this doesn't mean that, since God created more empty space, he loves empty space more than matter. In fact, the Abrahamic religions seem to put the ladder in the opposite order. Why? Because each step down *adds* something. Sure, there's more empty space than matter, but occupied space has a trait that empty space lacks. There's more nonliving matter than living matter, but living matter has a trait that nonliving matter lacks. And there's more nonrational life than rational life, but rational life has a trait that nonrational life lacks.[5]

The second source of value in Judaism and Christianity (but not Islam) is being created in the image of God, and this is what sets humanity apart from the rest of what he has created (ignoring the possibility of extraterrestrial intelligence). Unlike other things God has created and loves, humanity has a direct connection to God because we are created in his image, and this means that we have *greater* value than those things that lack this connection. This creates a problem, however: I've been

---

5. I'm not addressing the philosophical issue of whether these traits amount to distinct things, since many people would say that life and rationality are reducible to nonliving matter. Well, maybe not many *people*, but many philosophers, at least.

arguing that premodern people did not think human beings were valuable and important, and so claims to the contrary (based on the Myths of Dethronement and the Big Fish in a Small Pond) are dead on arrival. But now I'm saying the Jews and Christians among them *did* think human beings were important by virtue of being created in God's image. What means this?

III. IMAGO DEI

So, again, we are created by and in the image of the ground of all value, and this means that we have objective value. But what does this even mean? Lots of ink has been spilled on this, but I've dumbed this issue down for myself too. The following isn't even trying to be exhaustive, it's only making some points that are particularly relevant to the present discussion.

In Judaism and Christianity God is the ground of rationality and morality. Human beings have rational and moral senses; we are rational and moral *agents*. So, by being rational and moral we are participating in the divine nature, we are directly connected to the ground of all existence. We may not be *aware* of it, of course, but we are. Other elements of creation, excluding rational extraterrestrials if any there be, supposedly do not have these senses and so do not have this direct connection. Other animals don't have the capacity for rationality or morality. Cats don't *murder* mice, they just kill them. They're not moral agents, they're not able to participate in the divine nature as we are. There's certainly more to the concept of the image of God than this, but not less. Augustine and Aquinas, to take two obscure and marginal voices in the history of Christian thought, say explicitly that it is the rational and intellectual aspect of human beings that most constitutes their having been created in the image of God.[6]

Of course, the image of God is not exhausted by having a capacity for rationality. Traditionally, this doctrine tended to be expressed in a structural sense: having rational cognitive faculties meant being created in the image of God. But more recent analyses suggest we should understand it in a functional or relational sense.[7] In response, I will make three points:

---

6. Augustine, *On the Trinity* 14.6–11; Aquinas, *Summa Theologica* I, Q93, A2, A4, A6.

7. Vainio, *Cosmology in Theological Perspective*, 143–55.

1. A structural interpretation does not exclude functional and relational interpretations. Indeed, Aquinas's take explicitly includes all these aspects.[8]

2. Taking a functional or relational interpretation as definitive does not in any way preclude rationality as an aspect of the image of God. If a capacity for rationality has sometimes been overemphasized, this does not justify omitting it.

3. The reason rationality is included in the image of God is because it is an expression of God's nature. Thus, by being rational we are participating in the divine nature, and so are created in God's image (the same goes for morality).[9]

So, according to the Bible, we do have value that exceeds that of the other things that God created. Clearly, premodern folk thought that human beings were more important than the other elements of creation. There's an important and very obvious qualification though. The grandeur that attaches to the image of God comes from the fact that it is the image of *God*. The image bearers do not contribute anything to it. So whatever value human beings have is derivative. Nor does this suggest that God chose to make us in his image because we deserve it, as if we were just that great. A consistent biblical theme is that God usually chooses the lowliest things to be the vehicles of his revelation and grace. "Lowliest," here, can have two meanings: First is worst ("It is not the healthy who need a doctor, but the sick. I have not come to call the righteous, but sinners" [Mark 2:17]). The apostle Paul applied this to himself: "Christ Jesus came into the world to save sinners—of whom I am the worst. But for that very reason I was shown mercy so that in me, the worst of sinners, Christ Jesus might display his immense patience as an example for those who would believe in him and receive eternal life" (1 Tim 1:15–16).

The second meaning of "lowliest" is the least significant (1 Cor 1:26–29). God often uses the marginalized to channel his universal goals. For example, he chose the Hebrews to be his people despite their utter insignificance on the world stage at that time. And while we may approve of giving those who have hitherto been ignored their just due, we tend to be hostile to the idea that God gives a particular group special or unique

---

8. Vainio, *Cosmology in Theological Perspective*, 146–50.

9. A good summary of the different Christian approaches to this doctrine through the centuries is Peppiatt, *Imago Dei*.

access to him and knowledge of reality, even if it is a marginalized group. Whether we like it or not, however, is irrelevant.

> The mention of that nation turns our attention to one of those features in the Christian story which is repulsive to the modern mind. To be quite frank, we do not at all like the idea of a "chosen people." Democrats by birth and education, we should prefer to think that all nations and individuals start level in the search for God, or even that all religions are equally true. It must be admitted at once that Christianity makes no concessions to this point of view. It does not tell of a human search for God at all, but of something done by God for, to, and about, Man. And the way in which it is done is selective, undemocratic, to the highest degree. After the knowledge of God had been universally lost or obscured, one man from the whole earth (Abraham) is picked out. He is separated (miserably enough, we may suppose) from his natural surroundings, sent into a strange country and made the ancestor of a nation who are to carry the knowledge of the true God. Within this nation there is further selection: some die in the desert, some remain behind in Babylon. There is further selection still. The process grows narrower and narrower, sharpens at last into one small bright point like the head of a spear. It is a Jewish girl at her prayers. All humanity (so far as concerns its redemption) has narrowed to that.[10]

So what do we do with all this? God loves us and created us in his image. This means we are objectively valuable and important. But what's valuable and important about us is entirely about God, not us. God loves us but not because of anything positive on our part, as if we deserve it. He does not love us because we are lovable, he loves us because he is *loving*—love itself, in fact (1 John 4:8). But still, this love means human beings are objectively valuable, and being created in his image means that we can participate in God's nature (by being rational and moral). Then add to the mix our utter insignificance in premodern cosmology by virtue of being a tiny speck in the most disreputable neighborhood in the universe.

These combine to form a complex picture. On the value spectrum, we're at both ends simultaneously. Pascal put it best: "What a chimera then is man! What a surprise, what a monster, what chaos, what a subject of contradiction, what a prodigy! Judge of all things, weak earthworm;

---

10. C. S. Lewis, *Miracles*, 120.

repository of truth, sink of uncertainty and error; glory and garbage of the universe!"[11]

In contrast, the naturalist picture says that human beings don't have value, but they don't have any disvalue either. We're in the middle of the spectrum. But that's because there's nowhere else to be: *everything's* in the middle of the spectrum (which effectively means that there is no spectrum in the first place). There's no source of value, but that means that nothing is either wonderful or horrible. It just is. At least I *hope* that's what they say because otherwise they would be saying that the earth not being at the center *does* indicate it's not important, with the corollary that whatever is at the center of the universe truly *is* the most important thing. Also that the biggest thing in the universe is the most important thing. But, as I say, naturalists aren't arguing this. They think they're correcting the premodern indicators of value and refuting them on their own terms, but the subtext is that those indicators didn't actually indicate value in the first place. Yet, as we've seen, the premoderns didn't think geocentrism and the universe's size indicated value. They indicated either disvalue or insignificance. What indicated value to them was being loved by God and being created in his image. *And how has science challenged this at all?* It hasn't. Science says nothing about how important we are to God, how much he loves us, whether we have rational and moral senses, etc.

Besides which, what's the alternative? If human beings don't have value, then there isn't anything morally wrong with hurting other people. Most of us wouldn't do it anyway because we don't have any particular *desire* to harm others. But if people don't have objective value, then there's nothing wrong, really wrong, with moral atrocities, like the killing fields of Cambodia or the rape of Nanjing. Most of us, if not all, are more confident in calling these things immoral than we are in any metaphysical theory that challenges it. Nor can we say, since the Judeo-Christian worldview puts people at both ends of the spectrum simultaneously and the naturalist worldview puts people (and everything else) in the middle, that they balance out. They don't. These are two different visions of human value.

Now I have to add a very important caveat here: simply being a naturalist doesn't mean that you don't (or shouldn't) believe that human beings have value. There are plenty of first-rate philosophers who are naturalists *and* moral realists—that is, they believe moral truths are objectively true. This may create a tension in their philosophy (at least,

---

11. Pascal, *Pensées*, 36. I love that phrase, "glory and garbage of the universe." It would make a great tattoo.

according to their critics), but that's another issue. On the other hand, some people have used naturalism or atheism as justification to commit horrific atrocities, just as people have often used belief in God to commit atrocities. Some serial killers, like Jeffrey Dahmer and Israel Keyes, stated that they didn't think human beings had value, and that this was one of the factors spurring their crimes.[12] During the Cold War, communist torturers would appeal to the nonexistence of God to deny that their victims had any value or that torturing them was morally wrong.[13]

But none of that is *science*. Science has nothing to say about whether human beings are loved by God or are created in his image. What scientific discovery shows anything that even touches this?

IV. EVOLUTION AND MORTIFICATION

Well, some answer, *evolution*. We thought we were created by God in his image but it turns out we're just animals. When Darwin first suggested his theory, Huxley and others took it in this direction. To make the case, they pointed to other examples of science mortifying vain human pretensions to value (such as heliocentrism and the universe's size) to show that there's a pattern of religion saying people have value and science saying no they don't. *But there is no pattern*. Decentralizing the earth and the universe's size do not show that we do not have value because people thought they reflected a lack of value. But even so, people also believed that the real indicators of value had to do with our connection(s) to God, and none of these issues touch that.

Whether there's a pattern or not, though, does evolution mean that we're merely animals and that we're not created in the image of God? No, it doesn't. It shows that we're animals, but everyone already believed that. The traditional definition of human beings—from Aristotle, again—is that we are rational animals. People knew perfectly well that there were obvious similarities between human morphology and that of other animals.[14]

---

12. Dahmer said in an interview in prison that he *did* believe that, but that he had since come to believe in God. Keyes wrote about one of his victims, "Soon, now, you'll join those ranks of dead or your ashes the wind will soon blow. Family and friends will shed a few tears, pretend it's off to heaven you go. But the reality is you were just bones and meat, and with your brain died also your soul." See his note at https://www.dailymail.co.uk/news/article-2274650/Israel-Keyes-The-chilling-page-note-body-Alaska-serial-killer.html.

13. Wurmbrand, *Tortured for Christ*, 38.

14. Ferngren, *Medicine and Religion*, 101.

We're not less than animals, but that doesn't mean that we're not more; and the "more" in question is Aristotle's adjective: we're *rational*. Remember, Augustine and Aquinas thought intellect was the primary aspect of being created in God's image. If we have that property or potential, then it doesn't matter whether it happened by natural or supernatural processes. If we have a rational sense, we're created in God's image.

Evolution is supposed to be mortifying because it shows that we're descended from animals that are completely amoral. Many animals procreate by force and eat by violently killing and consuming members of their own species. Such actions would be deeply immoral if human beings did them. Or maybe people are just disgusted by the idea that we are descended from creatures that fling their poo. I don't know. I have heard Christians acting all indignant at the idea, but they aren't representative and all they can do is appeal to emotion. What exactly would the problem be? Sure, our distant ancestors didn't have moral or rational senses. So what? We're *descended* from them, but we *aren't* them. We are animals who *do* have moral and rational senses; we even occasionally express them. This is like saying we should be humiliated by the fact that we all started off as babies who crapped themselves. Or perhaps even flung their poo.

In fact, the Bible says that God made the first human beings out of mud (Gen. 2:7). That's a more mortifying, humiliating, dishonorable origin than to be descended from animals. After all, animals have life, breath, blood, and soul,[15] which in the Bible are all indicative of value. So which is more mortifying? Being created from creatures with life, breath, blood, and soul, or being created out of mud? Of course, this is academic since animals were originally created out of mud too; evolution just gives some of the intermediate steps between mud and human beings.

Maybe the mortification isn't that human beings were created from animals but that we're *not* created by God, and thus not created in his image. But evolution doesn't say we're not created by God, only evolution + atheism does. We have to take evolution as a *competing account* of our origins in order for it to suggest that we're not created by God and therefore (poor inference) not loved by him and not important to him. But to take evolution as a competing account is simply to wed it to atheism from the outset, or at least with the denial of the Abrahamic religions. Of course, at that point, it *would* be incompatible with these religions,

---

15. Yes, soul. That's the most common word used in the Old Testament to refer to animals, "soul" (*nephesh* in Hebrew).

for the same reason that merging Mendelssohn's Violin Concerto in E Minor with the denial of Abrahamic religions would be incompatible with the Abrahamic religions. Anything tied to the denial of Christianity is incompatible with Christianity, but that doesn't make the thing so tied incompatible. All the evidence for evolution is equally consistent with evolution + theism and evolution + Christianity as it is with evolution + atheism (or, for that matter, evolution + moon-landing conspiracy theories). In order to choose between these options, we have to have some other reason beyond the bare fact of evolution itself. And the doctrine that human beings are created in God's image is not diminished by saying God created human beings in his image *via the evolutionary process*. It just takes the statement of who created us and adds to it a statement of how he did so.

Remember the four causes. Using those, theists would say that God is the efficient cause of human beings and evolutionary processes are, or describe, the material causes. Naturalists want to say that the evolutionary processes are the efficient causes (or maybe the exemplary cause: these processes being a description of how the efficient causes work). But the science doesn't commit us to either suggestion or make one more plausible than the other.

V. OTHER MORTIFYING DISCOVERIES

I've gone over the big three, the primary ways by which it is suggested science has mortified or humbled human beings: decentralization, the idea that we're not at the center of the universe; the size of the universe, which shows how spatially irrelevant we are; and evolution, which shows that we are descended from animals, which were descended from other animals, all the way back to an original, single-cellular form of life. And I've pointed out that none of these is mortifying or humiliating. Decentralization doesn't do it, because people thought being at the center of the universe reflected a *lack* of value; the universe's size doesn't do it, because people always thought that the universe was larger than we can imagine and that human beings make up a tiny speck within it; and evolution doesn't do it, because it doesn't show that we're not created in God's image unless you add atheism to it—in which case, it's already inconsistent with Judeo-Christianity on more obvious grounds.

I've also mentioned a few other points: motion was a good thing, and our apparent lack of motion, by being at the center, indicated that we weren't good; worse than everything else in fact, since everything else *is* moving. People also thought that everything but the earth shone, and again, since light is a good thing, this indicated that we weren't good. And the medieval T-O maps of the known world portrayed Europe as located furthest from paradise, Christ, and salvation.[16] Obviously all this must be weighed against the doctrine of *imago Dei*, that we are created in the image of God, but this suggests we're at both ends of the scale at the same time, not just at the good end. Sigmund Freud even suggested his psychoanalysis should be accepted as Real! Science! specifically because it fits into this alleged mortifying pattern: not only are human beings mere animals, we're *sick* animals, we're *malfunctioning* animals.[17]

There's one more potentially mortifying discovery, "potentially" not only because it could potentially mortify us but also because we haven't made the discovery yet: the existence of intelligent extraterrestrial life. This isn't fully addressed by the contents of this chapter because it adds an element to the already-complex picture of occupying both ends of the value spectrum simultaneously: Namely, what would be the spiritual status of intelligent extraterrestrials? What would be their relationship to God and (for Christians) Jesus? Is Jesus the savior of humanity or the savior of all creatures created in his image—and if the latter is the case, wouldn't that commit us to think that earth is important by virtue of being the planet where God took on the form of one of his creatures? Wouldn't it commit us to think that humanity is important by virtue of being the people God chose to be incarnated as? "Although being lowest in value, the earth was still the epicenter of the cosmic drama, the dwelling of humans, and witness to God's involvement with creation. Our home was provincial, but not forgotten. We were living in a galactic Bethlehem."[18] But if there are extraterrestrials out there, would earth still be Bethlehem? This deserves its own chapter.

---

16. Goldie, *Idea of the Antipodes*, 42.
17. Freud, "Difficulty in the Path of Psychoanalysis."
18. Vainio, *Cosmology in Theological Perspective*, 52.

Chapter 11

Xenomorphobia

DENNIS DANIELSON, AN EXPERT on early modern scientific literature, has been at the forefront of correcting the Myth of Dethronement. In an unpublished essay, "Copernicus and the Tale of the Pale Blue Dot," he states that when he gives lectures about geocentrism and how it indicated a lack of value and significance to the premoderns, scientists often come up to him afterwards and tell him that his claim is interesting but couldn't possibly be true. For them, the geocentrism issue is a microcosm of the whole history of science. Challenging it disrupts their understanding of what science is and what they are doing. If science isn't an iconoclast—or if it's not breaking the *right* icons—their view of the world and of themselves is threatened. Naturally this isn't true of all scientists, just some.

Danielson concludes the essay with another response people have given him: "If intelligent life were discovered elsewhere in the universe, what would it do to your theology?"[1] Before giving his answer we should ask what exactly the issue is. This is another one of those areas where some people seem to think that just mentioning the subject constitutes a final, incontrovertible refutation of Christianity or religion in general. *Why* it would be so isn't stated plainly—or at least convincingly. This poses a problem. It makes the claim difficult to refute, not because it's a good argument but because it's not clear what the argument *is*. Any attempt to address it, therefore, runs the distinct possibility of attacking a straw man. Nevertheless, I shall soldier on.

1. Danielson, "Copernicus and the Tale of the Pale Blue Dot."

I. ALIEN INDUCTIONS

In chapter 6 I mentioned the plurality of worlds debate when discussing Giordano Bruno. This is an old idea going back to the pre-Socratics. The atomists, including the Epicureans, maintained that all possibilities would be actualized in an infinite void with infinite atoms randomly forming into infinite configurations, which entailed that there were infinite worlds with an infinite diversity of infinite creatures on them.[2] In the twentieth century, Arthur Lovejoy ascribed to Platonism a "principle of plenitude," according to which God would not allow any potential to remain unactualized.[3] Thus, if other worlds could *potentially* exist, they would *actually* exist. This is a similar intuition behind the idea that a huge universe is "an awful waste of space" if there are no extraterrestrials.[4] But with Aristotle's physics, this wasn't possible. Objects want to go to their places: earth, air, fire, and water are heavy and fall down because that is their place, and in a spherical universe down is centerward. Asking if there could be more than one earth made as much sense as asking if a sphere could have more than one center.

Within early Christianity, Origen said God could very well have created other worlds with other creatures on them. A millennium later, Albertus Magnus, Aquinas's teacher, considered the plurality of worlds to be "one of the most wondrous and noble questions in Nature," indicating that it was not a forgotten issue, but ultimately rejected it. St. Bonaventure argued that God could certainly create other worlds if he wanted to. Some speculated about the moon and planets being populated with rational animals, but this was pretty limited compared to the contemporary question of intelligent extraterrestrials. In the buildup to the Scientific Revolution, it was seriously considered by other theologians like Nicole Oresme and Nicholas of Cusa, until, once geocentrism was called into question in the sixteenth century, it began growing into a more significant issue.[5]

---

2. Say "infinite" again.
3. Lovejoy, *Great Chain of Being*, 52–55.
4. See above, 66.
5. Origen, *De Principiis* 2.3.4; Dick, *Plurality of Worlds*, 23–43; McColley and Miller, "Saint Bonaventure," 387; Crowe, *Extraterrestrial Life Debate*.

II. WHAT'S NOT THE PROBLEM

Before anything else we have to ask whether it matters if whatever life we may discover is simple or advanced life. The Rare Earth Hypothesis is the idea that simple life may be common in the universe, since it is more robust. On earth we have even discovered extremophiles, simple organisms that thrive in (and require) conditions that are antithetical to the other life forms we know.[6] But the conditions that must be met for *advanced* life to exist are so particular and so uncommon that, according to the hypothesis, it's unlikely they will be met anywhere else in the universe.[7] And before you ask, yes, the scientists who say this are aware of how big the universe is.

The discovery of simple life elsewhere in the universe wouldn't pose a problem for religion in general or Christianity in particular, any more than the discovery of extremophiles in remote and toxic locations on earth does. What we're ultimately addressing is the possibility of *rational* life elsewhere in the universe—not ET but ETI (extraterrestrial intelligence). Part of the reason for this is that we may discover the remains of life on Mars, just because several million tons of our planet has been dumped on that planet over the last billion years or so, through meteor strikes propelling earth material out into the solar system. The odds that it has no biological material at all are remote in the extreme. Solar radiation would break it down over time, but some of the remains could still be present, at least on a microscopic level. Additionally, simple organisms in the upper atmosphere, which can survive for extended periods in vacuum, are pushed further out by solar wind.[8] Biological material from earth will probably have settled on other bodies in our solar system as well, such as the moons of the outer planets.

One reason why some people think the discovery of simple life elsewhere in the universe would be problematic is because they think it would show that life arises by natural processes (no God need apply). But even if it did, so what? That should pose no more of a difficulty than saying that life *diversifies* by natural processes, i.e., evolution. Neither of these is a problem since the God in the Abrahamic religions usually

---

6. Carré et al., "Relevance of Earth-Bound Extremophiles in the Search for Extraterrestrial Life."

7. Ward and Brownlee, *Rare Earth*.

8. Shklovskii and Sagan, *Intelligent Life in the Universe*, 207–11; Hoyle and Wickramasinghe, *Evolution from Space*, 39–61.

produces his effects via the natural processes that he set up in the first place. In fact, I suspect there is a natural origin of life just because when God creates plants and then animals in Gen 1, the text says he delegates it to the environment ("Let the land produce . . ." [vv. 11, 24]). This could apply to the origin of life, the evolution of species, or both. To be sure, it doesn't *require* a natural process rather than a supernatural one, but it sounds like it to me. However, if God tells me I'm wrong when I go before him, I won't argue the point.

Another reason the presence of simple extraterrestrial life could be problematic is that once simple life is established, some people think it's bound to evolve into complex life, and then to intelligent life.[9] But the whole point of the Rare Earth Hypothesis is that the conditions that would allow such an evolutionary process are so numerous, specific, and uncommon that they are unlikely to reoccur—at least by natural processes, which is an important qualification. The mere presence of simple life elsewhere in the universe does not entail that it will evolve into rational, intelligent life, so it is not relevant to the problems surrounding the possibility of extraterrestrial intelligence.

III. WHAT'S STILL NOT THE PROBLEM

One idea that's thrown around is that if we discovered rational extraterrestrials, they would have their own religions, which would imply that our religions are not true. How likely is it that the worldviews we've come up with should be cosmically applicable and relevant to creatures who may be much more advanced than us and have a much broader (or different) stockpile of information on which to base a worldview? But this assumes that religions are entirely, and so *merely*, a product of biological and cultural issues rather than revelations from a supernatural reality, so that we can't take them as objectively correct. But this is not given. If a religion is actually based on actual revelations from an actual God, then to dismiss it as a mere cultural phenomenon is a pretty significant misstep. It assumes that the claims of Christianity and all other religions are false in order to argue that they are false, and this blatantly begs the question.

Moreover, the suggestion that the discovery of extraterrestrials with different religions should have some relevance to whether a human religion is true is just the problem of religious particularism in the face of

---

9. Chela-Flores, "Phenomenon of the Eukaryotic Cell."

other religions: whether it's appropriate to uniquely affirm Christianity or Islam or Judaism when there are other religions out there. *This is already an issue.* And what makes it an issue is the mere *fact* of other religions, not their quantity or the diversity of their contents. Extraterrestrial religions would certainly be fascinating, but they would not pose a new problem for the Abrahamic religions. I don't mean to just dismiss it, but the treatment of this issue is the same whether we take potential extraterrestrial religions into account or not; throwing aliens into the mix doesn't add anything to it. Nor will I, since that's not the subject of this book.

IV. THE PROBLEM

Imago Dei, again

While I don't think the discovery of intelligent extraterrestrials can be made into an argument against Christianity or theism or religion, I do think it poses a problem insofar as it would force us to ask difficult questions. The first is whether Jews and Christians should think that extraterrestrials are created in God's image as human beings are. Yes, as noted in chapter 3, Christianity is not anthropocentric so much as "anthropo-peripheral."[10] Humanity is not the center of importance, God is. And in a theocentric cosmos, God will have some relationship with whatever intelligent extraterrestrials there may be.

One thing to say in response is that, since they would be, ex hypothesi,[11] intelligent creatures displaying rationality, they would be exhibiting one of the primary elements of what being created in God's image has traditionally meant.[12] Perhaps someone could object that they might exhibit apparent rationality without really being rational, but this just boils down to the problem of other minds: how we know that other people are individual loci of self-consciousness and first-person perspectives.[13] Granted, it would probably pose a bigger problem when the other minds in question do not resemble us in other ways, since I suspect the intuition wouldn't be felt as strongly. But some people talk to their plants or cars as if they were people, so I don't think this would be insurmountable.

10. C. S. Lewis, *Discarded Image*, 58.
11. From the Latin *ex* (never) and *hypothesi* (gonna give you up).
12. For a counterexample from science fiction, read Blish, *Case of Conscience*.
13. *How* we know it, not *whether* we know it. We can trust our intuition that other people are minds, but how did we come by this knowledge?

Anyway, as with the problem of religious particularism, the problem of other minds is already an issue so including extraterrestrials doesn't add anything to it.[14]

Original sin and salvation

If we accept that ETIs are created in the image of God by virtue of being rational, we have to ask further questions, and these we could not answer, absent further revelation. Foremost of these is whether they are fallen as humanity is; whether they have the stain of original sin. This is the claim that we are born with an innate tendency to sin as a consequence of our distant ancestors sinning against God. This is a difficult doctrine since it seems to imply that we are either being punished for the actions of someone else or we're being punished for being born into a state of sinfulness. Neither of these is under our control, though, so how can we be held guilty for them? How could it be just to hold us accountable for things we had nothing to do with? In fact, the Bible explicitly says that holding people responsible for something their ancestors did is not just—that while the *sins* of the father may be visited upon the sons (Exod 20:5),[15] the *punishment* for the sins of the father will not be, at least not by God (Ezek 18).[16]

As is often the case, C. S. Lewis makes this doctrine very understandable:

> Our present condition, then, is explained by the fact that we are members of a spoiled species. I do not mean that our sufferings are a punishment for being what we cannot now help being nor that we are morally responsible for the rebellion of a remote ancestor. If, nonetheless, I call our present condition one of original Sin, and not merely one of original misfortune, that is because our actual religious experience does not allow us to regard it in any other way. Theoretically, I suppose, we might say "Yes: we behave like vermin, but then that is because we *are* vermin. And that, at any rate, is not our fault." But the fact that we are vermin, so far from being felt as an excuse, is a greater shame

---

14. Avramides, "Other Minds."

15. I take this to mean that children are very likely to adopt the bad behavior of their parents. Children of abusers often grow up to be abusers themselves, for example.

16. Some Bible translations fail to make this distinction but, in general, the more literal the translation, the more likely it is that it follows the original Hebrew text in distinguishing these two points.

and grief to us than any of the particular acts which it leads us to commit. The situation is not nearly so hard to understand as some people make out. It arises among human beings whenever a very badly brought up boy is introduced into a decent family. They rightly remind themselves that it is "not his own fault" that he is a bully, a coward, a tale-bearer and a liar. But, however it came there, his present character is nonetheless detestable. They not only hate it, but ought to hate it. They cannot love him for what he is, they can only try to turn him into what he is not. In the meantime, though the boy is most unfortunate in having been so brought up, you cannot quite call his character a "misfortune" as if he were one thing and his character another. It is he—he himself—who bullies and sneaks and likes doing it. And if he begins to mend he will inevitably feel shame and guilt at what he is just beginning to cease to be.[17]

The transmission of original sin

So if we encounter rational extraterrestrials and assume from their rationality that they are created in God's image, how would we be able to determine whether they are fallen? Not meeting our present culture's mores would not be much of a clue; it may even be a mark in their favor. But there are Christian concepts that make answering these questions even more difficult. The first, and more significant, of these is that, since original sin affects only human beings, it is transmitted "biologically," that is, from parents to children. This doesn't mean genetic per se, just that it's something inherent in the human condition since the fall.[18] Although the Bible does not explicitly make this point, it would explain why God was incarnated as a human being, viz., to atone for the sins of human beings. (Rom 5:18–19).

As such, the fall of humankind would not mean that any nonhuman rational extraterrestrials would be fallen as well. They could be unfallen. Or perhaps *their* ancient ancestors sinned against God as well, and they are fallen by virtue of that. If they are fallen, did their fall have the same or a similar effect as ours? And if it did, would Christ's atonement apply to them? Or was Christ incarnated among them too? Or do they have a completely distinct path to salvation? After all, different diseases require

---

17. C. S. Lewis, *Problem of Pain*, 85–86.
18. Harent, "Original Sin"; Beatrice, *Transmission of Sin*, 58–76.

different cures; even the *same* disease can require different cures among different people.[19] We may not even recognize a cure as a cure.

But *if* we were, somehow, able to determine that ETIs are fallen, what could the Christian conclude about their spiritual status? There are a few possibilities.

1. Jesus' atonement on earth applies to them. He atoned for all creatures created in his image, not just human beings. This is the Galactic Bethlehem Model. There is some great science fiction with this motif.[20]

2. Jesus was incarnated among them too and atoned for their sins independently.[21]

3. They are atoned for by some other method that doesn't involve incarnation and death on behalf of others.

4. God has not atoned for them *yet*. This could collapse into option 1, since perhaps God puts them in contact with us to provide the atonement Jesus offers to all rational creatures. On the other hand, it could just mean that God has not yet provided them with the atonement particular to them.[22]

All these options have theological weaknesses, some severe, as well as strengths.[23] And we could combine them: perhaps some ETIs are unfallen, some are saved by Jesus' atonement on earth, some are saved by Jesus being incarnated among them, and some are saved by some alternative method. Part of the problem is that Christians already disagree on the nature of the incarnation itself. One of the theological debates throughout Christian history is whether Jesus would have been incarnated as a human being if we had never fallen into sin.[24] How can we

---

19. "Religion and Rocketry," in C. S. Lewis, *World's Last Night*, 83–92 (esp. 87).

20. My favorite is Macleod, "Case of Consilience." This story takes its title from the book mentioned in n12 above.

21. See Vainio, *Cosmology in Theological Perspective*, 159–65.

22. Or they could be fallen but God has no intention of atoning for them. However, since the Bible says that God wants everyone to be saved (1 Tim 2:4; 2 Pet 3:9), this option would be available only if we presuppose that Christianity is false, so we can't use it to *argue* that Christianity is false.

23. See Peters et al., *Astrotheology*, pt. 4 (chs. 16–20). For a take on how other religions and worldviews might respond to intelligent extraterrestrials, see pt. 3 of the same (chs. 11–15).

24. Ted Peters, "One Incarnation or Many?" in Peters et al., *Astrotheology*, 286–300.

predict whether Jesus would be incarnated among ETIs when we can't even figure out what incarnation is *for*?

The real problem is that, if we encountered intelligent extraterrestrials, we would have no way of knowing which of these possibilities is the case. Without knowing this, we would not know what the appropriate Christian response should be. If God has given them their own method of atonement, particular to their condition (options 2 and 3), then it might be improper for Christians to share Christianity with them, since it would take them away from the atonement that God has provided for them. On the other hand, if Jesus' atonement on earth applies to them (options 1 and possibly 4), then it might be improper for Christians to *not* share Christianity with them. This would be a problem, but again, not a problem that makes Christianity any less plausible. It only means that the revelation received thus far does not provide us with the information needed to answer these questions.

If we deny that original sin is passed on biologically, we are left with no explanation of it. That would not be ideal, of course, but just saying we have not figured it out yet is not as devastating as some people seem to think. The Bible does have some interesting ideas where spiritual realities impact the material world, such as the idea that sexual intercourse unites a man and woman into one flesh (Gen 2:24; Matt 19:5; Mark 10:8; Eph 5:31–32). Perhaps some spiritual reality has some relevance as to how original sin (and salvation) applies to ETIs. "Our loyalty is due not to our species but to God. Those who are, or who can become, His sons, are our real brothers even if they have shells or tusks. It is spiritual, not biological, kinship that counts."[25] But this is speculative. If we discover a reason to reject the idea that original sin is passed on biologically, then we can regroup, but in the absence of any such reason, it is more rational to accept it.

The cosmic fall

The other Christian concept that makes this issue so intractable is that the fall of humanity also fundamentally changed the entire earth, possibly the entire universe—we are fallen creatures living in a fallen world. The other elements of creation may not experience original sin, as they are not moral agents and so lack the capacity to be sinners, but they are

---

25. "Religion and Rocketry," in C. S. Lewis, *World's Last Night*, 90–91.

still affected by the fall of humankind. Again, this point is not made explicitly in the Bible, but it is an honest attempt to understand it. So if we discover rational extraterrestrials, they may not be fallen by being under the specter of original sin, but they would be elements of a fallen world. This is not a contradiction—it would mean they are fallen in one sense and not fallen in another sense—but if they are rational and moral beings, it would imply they are created in the image of God, might not have sinned against him, yet suffer the consequences of *our* sin.

In an important sense, this is simply the difficulty of original sin. But if there are rational extraterrestrials, then the fall of one species created in God's image (us) ruined the universe that housed potentially unfallen species created in God's image *who had no biological connection to them*. Whatever injustice we feel with the standard view of original sin would be felt even more strongly with ETIs. Having said this, it still amounts only to an unanswered question, not an objection. It just means that we would need further information, further revelation, to resolve these issues.

Even though it is the traditional understanding, I think it is easier to challenge the idea of a cosmic fall than that original sin is passed on biologically. The passages where the consequences of the fall are mentioned in the most detail limit their context to human beings (Rom 5:12–21; 1 Cor 15:20–58), not applying them to animals or other forms of life, much less the nonliving world. Genesis 3:22–24 describes Adam and Eve's banishment from paradise, and the passage is often taken to suggest an alteration of the earth, if not the universe: "Cursed is the ground because of you; through painful toil you will eat food from it all the days of your life. It will produce thorns and thistles for you, and you will eat the plants of the field." Some suggest that thorns and thistles and other harmful forms of life were created at this point. In response—and leaving aside the question whether this text should be taken as historical—I contend that when Adam and Eve were expelled from the garden, God told them that the land would be cursed because they weren't going to be in the paradise God created for them anymore. When he says thorns and thistles would inhibit their efforts when they tried to plant and harvest crops, this is because they weren't going to be in the paradise God created for them anymore. This text refers to how they were going to live in the larger world that did not have the particular accommodations that Eden had. Thorns and thistles would have been created during God's creation week, after all.

Another biblical passage is often taken as blatantly describing a cosmic fall:

> For the creation was subjected to frustration, not by its own choice, but by the will of the one who subjected it, in hope that the creation itself will be liberated from its bondage to decay and brought into the glorious freedom of the children of God. We know that the whole creation has been groaning as in the pains of childbirth right up to the present time. (Rom 8:20–22)

Historically, many Christian theologians have taken this as a reference to the fall, although the text itself does not make that connection. The biggest problem I see with it is that the passage states that the one who subjected creation to futility did so with the intention of eventually freeing it. If we must assume that this has a connection to the fall of humankind in Gen 3, neither the human beings who sinned (Adam and Eve), nor the one who tempted them to sin (Satan), did so to "liberate" nature from its "bondage"; they sinned because they wanted to be like God (Gen 3:5). Besides these, the only other agent involved in the fall of humanity is God. Thus, the one who subjected the universe to futility "in hope that the creation itself will be liberated from its bondage to decay and brought into the glorious freedom of the children of God" must be God. But then why tie this to the fall? I take this passage as an anthropomorphic statement that God created the universe with certain limitations that will eventually be overturned with the creation of the new heavens and earth (Rev 21:1–5).

Another point brings the biological transmission of original sin into the mix. The passages that are used to defend this contrast it with the effects of the fall: "For if, by the trespass of the one man [Adam], death reigned through that one man, how much more will those who receive God's abundant provision of grace and of the gift of righteousness reign in life through the one man, Jesus Christ!" (Rom 5:17). But Christ didn't die to save frogs and slugs and chinchillas from death, much less nonliving elements of nature; he died to save human beings from death. But if the *atonement* applies only to human beings, the fall applies only to human beings. They covary with each other. And if we were to expand their application beyond humanity, we should expand it first to those elements of creation that could potentially sin and potentially be saved; in other words, to those created in God's image. That would apply to any ETIs there may be out there, but not to rocks, bugs, or jellyfish.

Basically, it comes down to whether Christ was incarnated as a human being to atone for human beings, or whether he was incarnated as a creature made in the image of God to atone for creatures made in the image of God. The former case would provide a solution to the transmission of original sin (that it's passed on biologically), but not a solution to the spiritual status of ETIs. The latter would provide a solution to the spiritual status of ETIs (the Galactic Bethlehem Model), but not a solution to the transmission of original sin. But in neither case do we get anything like an objection to Christianity, only an unanswered question.

In any event, I don't think that Christian theology needs to assert a cosmic fall, despite its being the traditional understanding. At the very least, it leaves plenty of wiggle room. And while this issue, along with the biological nature of original sin, makes knowing the spiritual status of ETIs more intractable, denying them doesn't exactly answer it either. But even so, not having answers to these questions does not impugn Christianity's credibility. Not being able to answer questions about the spiritual status of ETIs does not make Christianity any less plausible than does Christians' inability to ascertain, with any confidence, what incarnation is for or what the fate is of those who have never heard of Jesus. As Wolfhart Pannenberg writes,

> It is hard to see ... why the discovery of nonterrestrial intelligent beings should be shattering to Christian teaching. If there were such discoveries, they would, of course, pose the task of defining theologically the relation of such beings to the Logos incarnate in Jesus of Nazareth, and therefore to us. But ... their existence in no way affects the credibility of the Christian teaching that in Jesus of Nazareth the Logos who works throughout the universe became a man and thus gave to humanity and its history a key function in giving to all creation its unity and destiny.[26]

V. DÉJÀ VU

All the above should sound familiar. These are exactly the same issues that arose with the ancient and medieval issue of antipodes and antipodeans, people who were alleged to live on the other side of the world.[27] At the time, premoderns thought that it was impossible to get from one side to the other; they had no way of knowing if there was even *land* on the other side of the world. But if there was land, would it be inhabited by rational

---

26. Pannenberg, *Systematic Theology*, 76. This sounds like Pannenberg accepts the Galactic Bethlehem Model, but it doesn't require it.

27. See above, 21–23.

creatures created in God's image? If we can't get from hemisphere A to hemisphere B, any such creatures would have to be a distinct creation of God, not descended from the first human beings. And once this question is asked, it raises the same issues that rational extraterrestrials do. Are they created in God's image? If so, are they fallen? Are they saved by Jesus' atonement, or does that atonement apply only to descendants of those first humans? Do they have their own mode of atonement?

Thus, this issue is an ancient one. The questions are the same whether we ask them of antipodeans or ETIs. As such, we can take the answers the ancients and medievals gave to the former and apply them *mutatis mutandis* to the latter. Unfortunately, the answer they gave to the possibility of antipodeans was "How the #%$*& should we know?" Since the premoderns had no way of answering these questions, it was just easier for some (not all) to deny the existence of antipodeans to avoid having to speculate about their salvation or need thereof in the absence of any assurance or guidance (or, for that matter, motive). So, the issue has already been raised from time immemorial—*raised*, not *resolved*. I think the best lesson from this is to take a "cross that bridge when we come to it" approach, a lesson I am obviously ignoring.

Moreover, there's another similarity to extraterrestrials from the ancient world. This is the potential existence of so-called "monstrous races." These are the myths about creatures with one eye (cyclopes), with one giant foot (dufflepuds), with atrophied chests making their heads look bigger than they really are (intellectuals),[28] etc. Interestingly, since antipodeans were sometimes described as having their feet opposite ours, some took this to mean that their feet protruded behind them instead of in front of them, making them a monstrous race in their own right. These ideas did not originate within Christianity, but their potential existence raised the question of their spiritual status. Augustine argued that "we are not bound to believe all that we hear" about such creatures, but he did try to answer the status question. Basically, he wrote that if it's a "rational, mortal animal," then it's human and so has the same spiritual status as human beings, "no matter what unusual appearance he presents in color, movement, sound, nor how peculiar he is in some power, part, or quality of his nature."[29] However, he also took this to mean these races were descended from Adam and Eve: the idea of rational, mortal animals with distinct origins from humanity was ... *alien* to him (ba dum tss).

28. C. S. Lewis, *Abolition of Man*, pt. 1.
29. Augustine, *City of God* 16.8.

EXCURSUS: ARTIFICIAL IRRELEVANCE

These points might also apply to some concerns regarding artificial intelligence. This is an enormous subject and very controversial on numerous fronts. As it relates to religion, the big questions have to do with the nature of souls, minds, and materiality, but I think these issues are too complex to take on in a book of prolegomena. The only front I'll look at is whether AI poses a problem for religious beliefs in the same (or similar) form as do ETIs. After all, it would be another case of intelligent agents that are not biologically related to humanity. The differences, though, are substantial, not least that AIs would be technological rather than biological, and they *would* be related to us in a sense since we would have deliberately created them. So if AIs are genuine rational agents, leaving everything else aside, do these differences constitute a difficulty for religion and Christianity distinct from those of ETIs?

The first question would be whether AIs could be seen to have been created in the image of God if we are the ones who created them. Of course, creating new creatures in the image of God has been a common pastime throughout human history, but AIs would be a different nonbiological method of doing so. Would this somehow preclude the idea that God created them indirectly through us? I don't see how. Jews, Christians, and Muslims have always believed that God creates individual human beings at least partially through natural processes, and whatever supernatural processes they'd like to include in the procedure (like God giving the creature a soul at some point) would be just as available to AIs. So I don't think using a technological process rather than a biological one would somehow be beyond the pale. And, as I've argued, these religions leave plenty of room for God to have created humans in his image collectively via natural processes that he set up (abiogenesis and evolution), so I don't see why creating AIs in his image via humans would pose much of a problem. Since the image of God would probably apply to anything exhibiting rationality, it simply wouldn't matter whether God's involvement in its creation was natural, supernatural, or some admixture of both. If the suggestion is that AI researchers aren't trying to create anything in the image of God, we can respond by saying the same thing about atheists having children.

After this, we would have to ask whether AIs are fallen, and if so, in virtue of what? Is it enough that they were produced by fallen, polluted agents? Does original sin travel along technological pathways in addition to biological ones? Would they even have any awareness of a supreme being for them to rebel against or would we have to introduce the concept to them? And what would be their response to it? If we were somehow able to infer that they were fallen, would Christian salvation apply to them?[30] I mean, they aren't biologically related to us any more than extraterrestrials are—but they *are* related to us in a way ETIs are not. These are fascinating questions, but ultimately, I think we don't know the answers to them for the same reason we don't know the answers to the questions when we swap out AIs for ETIs: we don't have enough information to draw any conclusions with confidence.

Incidentally, there's an influential theorist in artificial intelligence named Jim Slagle. That's not me, it's *another* guy named Jim Slagle. I found his book in the library when I was doing my doctoral research and tried to submit it as my dissertation but it didn't work; partially because I was a toddler when it was published.[31]

VI. THE MYTH OF MORTIFICATION, AGAIN

If we say that the entire universe changed in some way when humanity fell it imputes an inordinate degree of power to human beings. This, supposedly, is one more example of human beings thinking we are more important than we are. If all the potential extraterrestrials out there were affected by the actions of a few human beings, then we have influence over them, an influence they do not have over us. Granted, in this case, the influence is bad, but still.

We already have the tools to resolve this. First, we could deny the fall had any effect on nature or we could limit its effect to the earth. Ignoring this, if there are rational extraterrestrials created in God's image who did not fall, it wouldn't mean they didn't go through a similar temptation, only that they didn't give in to it and so the universe was not changed. They would have had just as much opportunity and power to alter the universe as humanity, but made a better choice. Or perhaps they *did* fall and their original sins altered the universe too. Perhaps we're living in a universe that has experienced multiple falls by multiple species created in God's image. In a Cixin Liu novel, interstellar war involves altering the universe's laws and properties: e.g., you can slow down the speed of light (for defense

---

30. A *great* science-fiction story that touches on this is Jack McDevitt's "Gus."
31. James R. Slagle, *Artificial Intelligence*.

purposes) or remove a spatial dimension (for offense). Then it reveals that these measures have already been done—multiple times, in fact.[32]

However, there is no scientific evidence that the laws and properties of the universe have been altered in such a way and there *is* evidence they have not. But this, again, is already a problem for those who suggest there was a cosmic fall: the addition of other alterations from other cosmic falls from other creatures doesn't make it more of a problem.

The other side of this coin is the question: Would the discovery of rational extraterrestrial life mortify human beings, showing us that we're not as valuable as we thought we were? Again, I don't see how. We have value by virtue of being loved by God and being created in his image. The presence of other creatures that he loves and created in his image does not take away from that. God's love is unlimited. Extraterrestrials created in God's image presents no more of a problem in this regard than do other human beings created in God's image. We live in a world with over eight billion other people in it, all of them created in God's image, but this does not diminish the individual's value or God's love for them. How could it?

For my part, given that the God of the Bible clearly loves to create, I wouldn't be surprised at all if he created other intelligent life in the universe. For that matter, I wouldn't be surprised if he created other *universes*.[33] Really, I'd be a little surprised if he didn't. But we're not yet at the point where we can say one way or another. Clearly, I have some sympathy for the principle of plenitude.

What if we adopt the Galactic Bethlehem Model? Wouldn't that mean there's something more significant about humanity than ETIs, since God was incarnated as one of us and redeemed us first? But this assumes that God chooses the most important and valuable to be his representatives, and we've already seen he tends to do the opposite. If earth is the jump-off point for his cosmic redemption program, it means we're either the most depraved, the least significant, or both. Besides which, "If a thing is to begin at all, it must begin at some particular time and place; and any time and any place raises the question: 'Why just then and just there?'"[34]

Some elements of the Myth of Mortification use something to argue against God that, for most people, would point them *to* God. The unfathomable vastness of the universe, for example, triggers beliefs in

---

32. Liu, *Death's End*.

33. Another example from fiction: the "wood between the worlds" from *The Magician's Nephew*, C. S. Lewis's prequel to the Narnia books.

34. "The Seeing Eye," in C. S. Lewis, *Christian Reflections*, 175.

a prime reality that is much bigger than we are, and next to which we are as nothing. The potential discovery of extraterrestrials is another. According to surveys, religious people tend to say that the discovery of ETIs would *strengthen* their belief in God; it would produce a "God is amazing!" moment.[35] Nonreligious people may think that ETIs would or should pose a problem for belief in God or religion, but religious people disagree—or maybe they'd say that it might pose a problem for *other* religions, but not their own. In one sense, you'd expect them to say that, but I have met Christians who thought that if God created intelligent aliens he would have mentioned it in the Bible. This was one of the objections to antipodes as well: God wouldn't have allowed us to remain ignorant of their existence for so long. But as soon as you ask, "Why not?" the objection dissipates into nothing.

I started this chapter by referencing Dennis Danielson and how people ask him how the discovery of extraterrestrial intelligence would affect his theistic beliefs. I didn't give his response to this, but now I will:

> I believe the answer is that it would do the same thing that discovering intelligent life on earth does to my theology. It fills me with awe. It drives me to ask telic questions, questions about the purpose of life. It makes me ask who or what I am in relation to other intelligent beings. It fills me with gratitude that I can live in the same world, and share a moment in time, with other such contingent creatures. It fills me with wonder that there is something rather than nothing.[36]

This doesn't strike me as merely the right answer, but the answer that religious folk, specifically Christians, tend to think.

This shouldn't be surprising. Mortification *itself* is generally taken as pointing to God, regardless of how it does so. It produces that intuition of something vastly more important than ourselves that we should honor. To suggest that it points *away* from religion, that religion is primarily about how important *we* are, is simply ignorant. "It is a profound mistake to imagine that Christianity ever intended to dissipate the bewilderment and even the terror, the sense of our own nothingness, which come upon us when we think about the nature of things. It comes to intensify them. Without such sensations there is no religion."[37]

---

35. Peters and Froehlig, "Peters ETI Religious Crisis Survey."
36. Danielson, "Copernicus and the Tale of the Pale Blue Dot."
37. C. S. Lewis, *Miracles*, 55.

Chapter 12

A Brief History of Science and Religion

I. MYTHOS AND LOGOS

MANY ANCIENT CIVILIZATIONS MADE mathematical and scientific discoveries, but they were for utilitarian purposes. It wasn't until the ancient Greeks that people began studying these things for their own sakes. Thales (seventh–sixth centuries BC) is considered the first scientist and the first philosopher—not because he gave the right answers but because he asked the right questions.[1] Instead of explaining phenomena via stories (*mythos*) he tried to explain them via reasoning and observation (*logos*), exemplified today by philosophy and science, as well as other academic fields. Since the stories in question were often stories about the gods, some people take this as a move away from religion, thus polarizing science and religion and setting them against each other from the get-go.

But explaining things by stories is not intrinsically incompatible with explaining them by reason and observation.[2] We're a storytelling species. And the idea that *logos* is opposed to religion doesn't hold much water: in Christianity, Jesus is the cosmic *logos* (John 1:1–3, 14), the

---

1. Ring, *Beginning with the Pre-Socratics*, 19–22; Copleston, *History of Philosophy*, 1:22–24.

2. For a contrary (and radical) view, see Rosenberg, *How History Gets Things Wrong*. For a review of Rosenberg, see Van Woudenberg, "Self-Defeat, Inconsistency, and the Debunking of Science." For the value of stories in philosophy (not to mention all academic subjects and life in general) see Nina Rosenstand's excellent textbook *Moral of the Story*, chs. 1–2.

incarnated principle of rationality, and he often makes his points with parables, i.e., stories. More than this though, the primary subject of *logos* over the millennia has been the existence and nature of God. More ink has been spilled on that subject than any other.[3] In fact, the first *logos*-based question Thales supposedly asked was why things tend to hold together, why the universe exhibits any degree of order and structure instead of just being chaos; and the first *logos*-based argument he gave to answer it concluded that change and diversity presuppose an underlying unchanging unity—and later philosophers claimed he meant this in a religious sense.[4] There were similar developments and arguments, so that "by the fifth century BC, most educated Greeks were monotheists."[5] I don't want to overstate my point: if nearly everything is explained one way, and then a new form of explanation comes to town, obviously a lot of the first type of explanations are going to go by the wayside. But the fact that people sometimes explained things with a misleading story when a rational explanation would have been better isn't an objection to explaining things by story in general, any more than a bad rational explanation ("rational" in that it's not explaining via story) is an objection to rational explanations in general. Much less is it an objection to religion.

II. ANCIENT CHRISTIANITY

The handmaid's tale

So what was the response of Christianity to the science of the day? David Lindberg, one of the greatest historians of science ever ever, argued that the early Christian theologians generally saw science and philosophy as a low priority.[6] What was high priority was the edification of the church (which was going through frequent persecutions), the development of theology, and the defense and spread of Christianity. Science and philosophy simply were not a major obligation and took a back seat to these other issues.

However, Lindberg follows up, not being a major obligation does not imply that science and philosophy were not obligations at all. They were low priority but not no priority. Basically, they thought science and philosophy

3. Free will is a distant second.
4. E.g., Aristotle, *On the Soul* 1.5 (411a); Cicero, *De natura deorum* 1.10.25.
5. Herrick, *Philosophy, Reasoned Belief, and Faith*, 9.
6. David C. Lindberg, "Myth 1," in Numbers, *Galileo Goes to Jail*, 8–18 (esp. 16).

had *extrinsic* value but no (or little) *intrinsic* value. They were tools, not end goals. Pointless speculations were just that: pointless. Their value came in how they could be used to elucidate the things that really mattered. This is the basis of what became the handmaiden doctrine, that science and philosophy served (were handmaidens to) the queen: theology.[7]

But this worked only if the handmaiden was legitimate. Twisting it to make it say what you want was unacceptable. Augustine put it strongly:

> Usually, even a non-Christian knows something about the earth, the heavens, and the other elements of this world, about the motion and orbit of the stars and even their size and relative positions, about the predictable eclipses of the sun and moon, the cycles of the years and the seasons, about the kinds of animals, shrubs, stones, and so forth, and this knowledge he holds to as being certain from reason and experience. Now, it is a disgraceful and dangerous thing for an infidel to hear a Christian, presumably giving the meaning of Holy Scripture, talking nonsense on these topics; and we should take all means to prevent such an embarrassing situation, in which people show up vast ignorance in a Christian and laugh it to scorn. The shame is not so much that an ignorant individual is derided, but that people outside the household of the faith think our sacred writers held such opinions, and, to the great loss of those for whose salvation we toil, the writers of our Scripture are criticized and rejected as unlearned men. If they find a Christian mistaken in a field which they themselves know well and hear him maintaining his foolish opinions about our books, how are they going to believe those books in matters concerning the resurrection of the dead, the hope of eternal life, and the kingdom of heaven, when they think their pages are full of falsehoods on facts which they themselves have learnt from experience and the light of reason? Reckless and incompetent expounders of holy Scripture bring untold trouble and sorrow on their wiser brethren when they are caught in one of their mischievous false opinions and are taken to task by those who are not bound by the authority of our sacred books. For then, to defend their utterly foolish and obviously untrue statements, they will try to call upon Holy Scripture for proof and even recite from memory many passages which they think support their position, although "they understand neither what they say nor the things about which they make assertion."[8]

---

7. Aquinas, *Summa Theologica* I, Q1, A5.
8. Augustine, *Literal Meaning of Genesis* 1.19 (1:42–43). The quotation at the end is

The ancient and medieval Christians thought all truth is God's truth, and so truth from science or philosophy or anywhere else should be embraced by the Christian. This led to Anselm's (eleventh century) famous dictum, "faith seeking understanding":[9] we are not obligated to wait until all our questions are answered before choosing to trust God, but the Christian can't misuse this trust as some sort of justification to stop pursuing knowledge.

Most ancient Christian theologians used non-Christian philosophy extensively, including natural philosophy, aka proto-science. They were especially partial to Platonism and Neoplatonism, but ideas from Aristotelianism, Stoicism, Pythagoreanism, and elsewhere were used as well.[10] Roughly, they took a moderately eclectic approach to philosophy and science, employing the parts they thought congenial to Christianity and arguing against the parts they thought were not.

This doesn't put the high premium on science that we do, but Lindberg makes another important point: *no other institution or cultural force in the ancient world was as encouraging towards science as was Christianity.*[11] Yes, it was a low priority; yes, it had only extrinsic value; but this was still a more positive take on science than virtually every other cultural force in antiquity.

Insiders and outliers

One objection people sometimes give to all this is that there are ancient Christian theologians, such as Tatian, Basil, and especially Tertullian, who took a very low view of Greek and Roman philosophy, including science. Tertullian asked, "What indeed has Athens to do with Jerusalem?," Athens standing for secular knowledge and Jerusalem standing for religious knowledge.[12] But his critics misunderstand him. Tertullian was fond of hyperbole and making provocative statements. Basically, he was a third-century troll. Also the founder of Western theology, but still a troll. And he was fully aware of the philosophies of his day, accepting some and employing them in his arguments. He had a particularly high view of

---

from 1 Tim 1:7.
    9. Anselm, *Proslogion*, prologue.
    10. Lindberg, "Myth 1," in Numbers, *Galileo Goes to Jail*, 11.
    11. Lindberg, "Myth 1," in Numbers, *Galileo Goes to Jail*, 17.
    12. Tertullian, *Prescription Against Heretics* 7.

Platonism.¹³ The point is that he was not sniping at secular learning from the outside; he was criticizing it from the inside. He was one of them. This is true of Tatian and Basil as well, they were all insiders to the Greek and Roman academic traditions.

Moreover, these three are the most extreme voices on the Christians-criticizing-secular-learning team. They are not representative of theologians in general in this regard but are fringe players, outliers. Besides which, *none of their criticisms even addressed science in the first place*, just philosophy and non-Christian theology. So even ignoring all the above, they weren't critical of the science of the day.¹⁴

Tertullian also made the very provocative statement *Credo quia absurdum* (I believe because it is absurd), seeming to imply that he was abandoning all logic by accepting Christianity. This would be a statement of blind faith or fideism, which is condemned within Christianity. Faith means trusting God, but this is compatible with, and even requires, reason to play a part in it.¹⁵

Now Tertullian didn't actually write *Credo quia absurdum*, he wrote something slightly different that doesn't have as strong an implication.¹⁶ And, again, he often wrote hyperbolically. Besides which, he gave *arguments* for Christianity, so he obviously wasn't decrying logic, reason, and argument. In fact, *Credo quia absurdum* is one of his pro-Christianity arguments. He's saying that the doctrines of Christianity are not the kind of ideas that people have on their own. When we think about God on our own steam, our ideas are simple. We don't make up doctrines like that God is one essence and three persons, the second of whom became a human being who prayed to the first one. Or that the incarnated second one was fully divine and fully human simultaneously, thus comprising two distinct natures, not mixed together into one nature. It's not absurd in the logical sense but in the sense that people don't think up ideas like that. So the fact that Christianity does have these, scare quotes, *absurd*, end scare quotes, ideas in it is evidence that it does not originate solely with human beings, it must come from God. I'm not defending this argument, I'm just saying that was Tertullian's point.¹⁷

---

13. Lindberg, "Myth 1," in Numbers, *Galileo Goes to Jail*, 12.
14. Lindberg, "Myth 1," in Numbers, *Galileo Goes to Jail*, 11–13.
15. See below, 229–30.
16. Tertullian, *On the Flesh of Christ* 5.
17. González, "Athens and Jerusalem Revisited"; Siemens, "Misquoting Tertullian to Anathematize Christianity."

III. MEDIEVAL CHRISTIANITY

Past Romana

The early Middle Ages are roughly the five hundred years after the fall of the Roman Empire—at least the western part of it. (The eastern part relocated the capital to Constantinople and is called the Byzantine Empire.) The collapse of the western part had two momentous consequences: first, a lot of the ancient Greek writings were lost to Western Europe. Second, Western Europe ceased to be unified, making widespread intellectual collaboration or even communication extremely difficult. Granted, there were intellectual periods like the Carolingian Renaissance, but it was not as unified and expansive as it was under Roman rule. Despite this, however, there were significant scientific advancements in the form of practical technology throughout the early Middle Ages. Nailed iron horseshoes, plowshares, horse collars, water and tidal mills, stirrups, etc., were invented or borrowed and improved upon in the centuries following the collapse of the empire. These developments may seem underwhelming to us but they had an immense impact on the advancement of Western civilization and led to a population explosion. "Well before AD 1000, the population far exceeded what it was when the continent had been ruled by Rome, and remained above that level even after the Black Death had killed a third of the inhabitants of Europe in the fourteenth century."[18]

Ignoring and exaggerating the role of Islam

The Byzantine Empire continued the Greek and Roman traditions, but they weren't the wide-ranging dominion the Roman Empire had been. That role was filled in the seventh century when Islam came on the scene. By conquering numerous cultures and acquiring their sciences and technologies, Islam took on the mantle of an expansive empire that allowed for widespread intellectual collaboration. When Christian Europe began reconquering Spain from the Muslims in the eleventh century, they discovered a lot of Islamic writings, along with those "lost" texts that had been translated from Greek into Arabic, so they began translating them from Arabic into Latin. Their primary focus was on scientific texts, with the Qur'an and Islamic theology placing a close second. "Many societies

---

18. Hannam, *God's Philosophers*, 17.

are slow to accept that they can learn something from their enemies. This was not the case with medieval Europeans."[19]

In the modern era, Islam's contributions were downplayed until fairly recently. One of my master's theses was on medieval Islamic philosophy, and I wanted to pursue this as my doctoral research, but I feared I wouldn't be accepted into the program if I did. The old attitude, which was almost passé at the time, was that the Muslims only held on to what the Greeks and Romans wrote, copying and writing commentaries on them, until they passed them on to Christian Europe where intellectual advancement could begin again. This is just outrageous. Granted, a lot of Islam's contributions did come in the form of commentaries on Aristotle and others, but not all: there was significant development in medical science in Islam during this period, for example.[20] Besides which, commentaries are a valid medium for intellectual advancement. Read Ibn Rushd's (Averroes's) three commentaries on Aristotle's *De anima* if you doubt this.[21] I'm not saying he developed it in the right direction: in the long commentary he concluded that there must be an immaterial analogue to matter. This is completely insane—brilliant, but insane—thus demonstrating the significant overlap between philosophers and lunatics. Nevertheless, it's still a huge intellectual development. There was a school of Latin Averroists who were basically medieval Christian fanboys of Ibn Rushd.[22]

The same goes for Islam's role in passing on the ancient texts to Western Christianity. I have a book, *Scribes and Scholars*, that addresses how the medieval Christians got a hold of ancient Greek and Latin texts that spends a total of about a page (out of well over two hundred) on Muslim contributions. I don't mean to criticize it, it's an outstanding book, you just have to supplement it with something that addresses Islam's role.[23]

Unfortunately, the counterresponse is a huge overcorrection, just as outrageous and biased as what it's trying to remedy. There are so many superficial books out there painting Islam and Christianity one-dimensionally, claiming that Islam accomplished everything and Christianity wrecked it all.[24] That's just silly. Fortunately, Christians and Muslims al-

19. Hannam, *God's Philosophers*, 69.
20. Ferngren, *Medicine and Religion*, 125–27.
21. Ivry, "Averroes' Three Commentaries on *De anima*."
22. Copleston, *History of Philosophy*, 2:435–41.
23. Perhaps with Lindberg, "Transmission of Greek and Arabic Learning."
24. I was going to provide some references for this but it seemed mean-spirited to say, "Here are some idiots and their idiot books that say idiot things." I'm sure there's a

ready have plenty of other ways to denigrate each other without bringing their respective contributions to science into it.

The university

The High and the late Middle Ages are roughly AD 1000–1250 and 1250–1500 respectively. Starting in eleventh-century Europe, universities became a thing, which was obviously one of the greatest intellectual developments in human history.[25] Some people try to downplay it by suggesting that the only subject studied was theology, but this is incorrect.

> Most students never got close to meeting the requirements for studying theology (usually a master of arts degree). They remained in the faculties of arts, where they studied only nonreligious subjects, including logic, natural philosophy [science!], and the mathematical sciences. In fact, as a result of quarrels between faculties, students in the arts faculty were not allowed to treat theological subjects. In short, most students had no theological or biblical studies at all. Moreover, not all universities had a faculty of theology. Very few had one in the thirteenth century, and the newer foundations initially were not allowed to have one.[26]

Law was the most common subject in the medieval university, although about 30 percent of the overall curriculum was on scientific subjects.[27] Nor did students tend to be in the church hierarchy. At one point the Catholic Church decided to give university students clerical status,[28] the lowest rung of the ladder, but this was because a few of them had enraged the townies, who promptly gave them some local justice. Making them quasi-clerics meant they had to be tried in the ecclesiastical courts, which were *much* fairer and more evidence based than the local civil courts. Don't get me wrong, compared to our systems the ecclesiastical courts were pretty bad too, but if you were charged with a crime *at that time*, you would pray that it wouldn't be in a civil court.

---

more courteous way to express that, but it's eluding me at the moment.

25. Hannam, *God's Philosophers*, 74–76.
26. Michael H. Shank, "Myth 2," in Numbers, *Galileo Goes to Jail*, 19–27 (quotation from 22).
27. Shank, "Myth 2," in Numbers, *Galileo Goes to Jail*, 21, 23.
28. Shank, "Myth 2," in Numbers, *Galileo Goes to Jail*, 23–24.

From Paris with love

Starting in the mid-twelfth century Aristotle's works were reintroduced into Western Europe. They were polarizing since many traditional Christian doctrines had been forged in the fires of Plato and Augustine, and it wasn't clear you could bring Aristotle into the mix without causing serious problems. Many Christian theologians, however, immediately tried to work him into their systems. In the thirteenth century, Thomas Aquinas developed his theology in light of Aristotle and Ibn Rushd, both of whom were so uniquely important that Aquinas just called them "the Philosopher" and "the Commentator" respectively, as did others.[29]

In 1200, the various cathedral schools in Paris joined forces, and in 1215 officially became the University of Paris. But in 1210 a local synod had condemned Aristotle's works on natural philosophy—proto-science—forbidding students in the Arts Faculty from reading them because a relatively small group of heretics (the Amalricians) had claimed inspiration from them. Aristotle's logical and ethical works were still fair game though. It's very unlikely the synod had actually read any of the offending works. Five years later, this prohibition was written into the University of Paris's statutes. In issuing such a condemnation, Paris was extremely atypical.[30]

Its impact, however, was nominal. Since it applied only to Paris, students were free to switch to other universities where Aristotle was taught in his totality. In fact, the University of Toulouse advertised this availability of Aristotle in an attempt to seduce students away from Paris.[31] In 1231, the pope pardoned anyone who had read the texts (historians think they were being taught despite the ban) and allowed them to be read again: all they had to do was remove the supposedly offending passages. "We do not know if a committee was ever formed to expunge Aristotle's books because, as it turned out, no one bothered to remove anything."[32] By 1255 the previously-forbidden-but-not-anymore Aristotelian texts were required for degrees at Paris.[33]

29. E.g., *Summa Theologica* I, Q3, A5.

30. Hannam, *God's Philosophers*, 79–82; Shank, "Myth 2," in Numbers, *Galileo Goes to Jail*, 24–25.

31. Hannam, *God's Philosophers*, 81; Shank, "Myth 2," in Numbers, *Galileo Goes to Jail*, 25.

32. Hannam, *God's Philosophers*, 82.

33. Hannam, *God's Philosophers*, 82; Shank, "Myth 2," in Numbers, *Galileo Goes to Jail*, 24–25.

Aquinas was born in 1225. He went to Paris in 1245 to study under Albertus Magnus, then followed him to Cologne a few years later, and returned to Paris in 1252, first as a student, then as a professor. Afterwards, he taught at several universities, returning again to Paris in 1268. He passed away in 1274, not even fifty years old, but having effectively baptized Aristotle's entire oeuvre.

Three years later, 217 propositions were condemned in Paris, their focus being on Aristotle, Ibn Rushd, and those dreaded Latin Averroists mentioned earlier.[34] For the most part, the condemned propositions were that God's power was limited by natural laws, which was contrary to Christian theology. The person in charge of the condemnations was the intemperate Étienne Tempier, and he did so by taking permission of the pope to investigate the matters as a license to excommunicate. Once again, however, it applied only to the University of Paris, and once again, it didn't take, not least because Aquinas's philosophy had shown the great value of Aristotle and Ibn Rushd in the exposition and defense of Christian theology. Interestingly, the condemnations of 1277 ended up being one of the most positive developments leading to the Scientific Revolution a few centuries later since they allowed scholars to speculate beyond the framework of Aristotelianism and other philosophies.[35] For example, Plato and Aristotle argued that the concept of other worlds—which could mean other earths or other universes—was nonsensical; and given their philosophies, it was. But one of the propositions Tempier condemned was that God cannot create other worlds, and this played a role in reopening the issue of the "plurality of worlds" hypothesis as discussed in chapter 11. Ultimately the condemnations were overturned a half century later for the most part, but by then people were all a-speculatin'.

IV. LET THERE BE LIGHT

Now we move from the "Dark Ages" to the "Renaissance" and "Enlightenment," three outstanding terms of propaganda. The latter, along with the Scientific Revolution, are the most significant markers of the modern age in which we currently live. It's awkward to call a specific period "the modern age" because moving beyond it ends up sounding confusing. We

---

34. Thijssen, "Condemnation of 1277."

35. Edward Grant, "Science and Theology in the Middle Ages," in Lindberg and Numbers, *God and Nature*, 49–75 (esp. 54–59); Thijssen, "Condemnation of 1277."

have postmodernism, but that just sounds weird. It's like calling an era "the Current Age" and so the following era has to be called "the Aftercurrent" or something. "We live in the period after the period that's happening now."

The modern era began about 1500; the Scientific Revolution started in the mid-sixteenth century, and the Enlightenment took place in the "long" eighteenth century, from the late 1600s to the early 1800s. The Renaissance was a series of "rebirths" that took place at different times in different places throughout Europe between the fourteenth and seventeenth centuries. All these periods are often portrayed as a complete rejection of the theological nonsense of the Middle Ages, but there are some very strong continuities between them, and between medieval theology and early science in particular.[36] I don't mean to imply that the Scientific Revolution wasn't a revolution. There were plenty of things that medieval Christianity *didn't* contribute, such as the artificial experiment, which was one of the main factors in the development of modern science.

Two motives

As with any huge intellectual development, the Scientific Revolution was complex, involving many diverse, intertwined factors. But I'm going to pretend there were just two.

First was the desire to have control over nature, either altering it according to our wants and needs, or preventing it from altering us. This may sound like a hostile mindset, and sometimes it was,[37] but not always. Medicine falls entirely under this heading: it's the attempt to maximize health and wellness over against the impact nature has on our bodies. And technology is an attempt to make life more bearable. Certainly, this can be used in a negative way, but that just means such *usage* is bad, not technology itself. The representative here is Francis Bacon, the pioneer of the scientific method.

But science wasn't the only method people used in the early modern period to try to control nature. This was also the age of magic and witchcraft.[38] Of course, every culture has had magical concepts and

---

36. The influence of medieval theology on the Scientific Revolution is a much bigger story than I can tell here; just read Hannam's book *God's Philosophers*.

37. For an especially ugly example, see Sampson, *Six Modern Myths*, 77.

38. Thorndike, *History of Magic and Experimental Science*.

The cottage industry of ginning up conflicts between religion and science culminated in two enormously influential books: John William Draper's *History of the Conflict Between Science and Religion* (1874) and Andrew Dickson White's *A History of the Warfare of Science with Theology in Christendom* (1895). As is obvious from their titles, they strongly emphasized that there was hostility between science and religion, especially Christianity. To that end, they produced example after example demonstrating their theses. When you try to verify their claims, however, it doesn't end up in their favor. They weren't the only such books written around this time, not by a long shot, but they were the most influential. Nevertheless, they were recognized by historians of science as propaganda a long time ago. Despite this, they continue to be cited as reliable sources by those with an axe to grind.[46]

While we can't ultimately blame these two books for the Conflict Thesis, they amounted to pouring gasoline on an already-burning fire. The antireligious (not *non*religious) elements of society used their claims to create a barrier between science and religion so that accepting one effectively denied the other. After a few decades, large swaths of the Christian community acquiesced to this idea. I've already mentioned this but surrendering the terms of the debate to one's opponents is generally not a good idea, since it means you're letting the people who disagree with you tell you what your position is.

V. ON OVERSIMPLIFYING COMPLEXITIES

Throughout this book I've corrected urban legends that science and religion conflict, but as mentioned in the preface, my primary purpose has been how people try to pit science against religion, not how people try to pit religion against science. I have mentioned some of the latter, but since that wasn't my foremost goal, it makes the book a little one-sided. So, to be clear, I'm not dismissing the other side of the equation; it just isn't the issue I'm addressing.

Additionally, by correcting apparent conflicts I don't mean to imply there were never any tensions between science and religion. Sometimes there were. Sometimes they happily worked together. Sometimes they

---

46. Ronald L. Numbers, "Introduction," in Numbers, *Galileo Goes to Jail*, 1–3, 6. As an example of axe grinding, Herman Philipse approves White as "still of great value" despite acknowledging that many of his claims, including "his warfare-metaphor," do not stand up to scrutiny (*God in the Age of Science?*, 8n10).

coexisted without much of a glance at each other. Sometimes they offered each other mutual support. Lindberg writes, "If we speak of 'the Christian position,' we can only mean the 'center of gravity' of a distribution of Christian opinion, for great variety existed. The church was not monolithic, and there was no universal Christian view of pagan philosophy or natural science."[47]

In premodern interpretation, there were two essential principles of accommodating scientific truths within Christianity.

> In discussing questions of this kind two rules are to be observed, as Augustine teaches. The first is, to hold the truth of Scripture without wavering. The second is that since Holy Scripture can be explained in a multiplicity of senses, one should adhere to a particular explanation only in such measure as to be ready to abandon it if it be proved with certainty to be false, lest Holy Scripture be exposed to the ridicule of unbelievers, and obstacles be placed to their believing.[48]

For Christians, this was especially true of the creation account.[49] Generally speaking, they did not take the Bible as a sort of *Encyclopedia Cosmologica*, explaining everything there was to know about the world. Yes, some Christians did, and do, suggest that some things (like antipodeans or extraterrestrials) would have been mentioned in the Bible if they existed. Nevertheless, "the notion that any serious [premodern] thinker would even have attempted to formulate a world view from the Bible alone is ludicrous."[50]

I don't want to overplay my hand here: the diversity between Christians and Christian communities sometimes involved an angry and off-putting dogmatism—not as bad as contemporary American politics, but still. My point is just that the two fundamental principles of interpreting the Bible in the light of scientific claims did not conflict with each other. Holding the Bible as unassailable did not lead them to reject scientific claims unthinkingly. Remember general revelation: in Christianity, the

---

47. David C. Lindberg, "Science and the Early Church," in Lindberg and Numbers, *God and Nature*, 19–48 (quotation from 40).

48. Aquinas, *Summa Theologica* I, Q68, A1; Augustine, *Literal Meaning of Genesis* 1.18–1.19, 1.21 (1:42–43).

49. Grant, "Science and Theology in the Middle Ages," in Lindberg and Numbers, *God and Nature*, 63.

50. Lindberg and Numbers, "Beyond War and Peace," 342.

universe itself is revelation from God and so cannot be (and, historically, has not been) ignored when describing reality.

So attitudes to how Christian faith interacted with science and reason have been diverse. This is at least partially because sometimes they seem to give different answers to the same questions. There are many, *many* cases where scientific discoveries are not even remotely what one would predict from the Bible alone. In one sense this is to be expected: when you have two different fields with two different approaches addressing the same subject, it's unlikely they'll happen upon the same answer. We date the origin of humanity differently depending on whether we base it on paleontology or genetics. This doesn't detract from either field.

This is all well and good, a critic may say, but there's a deeper element when we address science and religion. It's not merely a matter of two fields addressing the same subject, but two worldviews. But of course, that *assumes* that the two approaches are in competition with each other, which is just to reassert the Conflict Thesis. Moreover, it assumes that science is itself a worldview rather than a method (or series of methods) for discovering the truth about the physical world. But this isn't science, it's *scientism*, which claims that only science can be trusted to discover the truth—a statement that is not itself a scientific one and so, by its own lights, is not trustworthy.

However, science and religion certainly amount to predominant social phenomena, so when they provide different answers to the same questions, it is, of course, no small thing. The point is that the relationship between science and religion has been and is complex and cannot be characterized in a simple, univocal way—especially if the characterization is as one-dimensional as the term "conflict" suggests. For these and other reasons, historians of science don't call it the Conflict Thesis anymore: they call it the Conflict *Myth*.

Chapter 13

Miracle Panegyrical

I. A PRIMER ON MIRACLES

ONE OF THE MAIN difficulties people have with religion as it relates to science is miracles. "Miracle," here, doesn't refer to something that makes us think of God, like a beautiful sunset or a newborn baby—"Oh, it's so lovely, it's just a *miracle*, isn't it? Pass the bourbon." It refers to events that are not plausibly explained by natural processes. You know, like a dead guy being resurrected in a glorified body. That kind of thing.

One objection to miracles is that they contradict the laws of nature. Perhaps people could have believed in miracles in earlier times because they didn't know there were laws of nature, but now that we *do* know (science!), we can conclude that miracles just can't happen. The problem with this is that the concept of miracle is *dependent* on the idea that there are laws and regularities of nature; otherwise there couldn't be exceptions to them. People had to know that water doesn't turn into wine in and of itself to recognize Jesus' turning water into wine as a miracle. If not, why even take note of it? They had to know that people thrown into a furnace would burn up to recognize an exception to this as a miracle. They had to know that people can't walk on the surface of the water to recognize that Jesus doing it was not a natural event but a supernatural one.[1]

Certainly, we know *more* about the natural laws and regularities than they did. We know how a spermatozoon (a haploid) combines its genetic

---

1. C. S. Lewis, *Miracles*, 49–52.

Miracle Panegyrical

material with an ovum (another haploid) to produce a new organism (a diploid). People in the ancient world did not have this information; they only knew that the man put his junk in the woman's lady parts and yada yada yada out comes a baby nine months later. But they still knew there was a natural law about this so that if a woman who had never been within firing distance of any man and his junk got pregnant, it couldn't have come about by natural processes; it would have had to be a miracle. In comparison, our scientific knowledge of the processes involved show that, for such a woman to get pregnant . . . it couldn't have come about by natural processes; it would have had to be a miracle.[2]

II. MIRACLES AND THE LAWS OF NATURE

Some laws, like moral or civil laws, are prescriptive—that is, they prescribe what *should* happen but often doesn't. Physical laws, by contrast, are *descriptive*; they describe what happens. Granted, these can sometimes be phrased with a "should"—"The plums should be ripe by now"—but this is just prediction based on previous description, it's not really prescriptive, as if the plums were under some sort of obligation to be ripe.

Another aspect of physical laws is that they can be treated as absolute if no other forces are in play. They hold *ceteris paribus*, or all other things being equal. Unfortunately, all other things are rarely equal. We can try to isolate them from other forces to study their particular properties; for example, by testing gravity in a vacuum chamber so that air pressure doesn't influence it. In the real world, we don't experience gravity in isolation; we experience a potpourri of forces affecting each other, one sometimes overruling another. If we drop an iron ingot, gravity dictates that it will fall, unless another force, say, a magnet (in the right place and of the right strength) is in play. The magnet slows the rate of fall, may even arrest it, and make the object go up. Does this mean the magnetic force is contradicting gravity or violating it by not allowing it to operate? Of course not, it's simply superseding it in this instance. Gravity is still just as much in force; there's simply another force that is stronger than it in that circumstance.

Science tells us about the regularities in nature. It doesn't say anything about *exceptions* to these regularities. Actually, "exceptions" is the

---

2. "Religion and Science," in C. S. Lewis, *God in the Dock*, 62–75 (esp. 66–70); C. S. Lewis, *Miracles*, 50–51.

wrong word, since it implies that something is being taken away—that a law that normally operates is being temporarily interrupted or suspended. But the idea of a miracle is that something is being *added*, not subtracted,[3] in the same way that the magnetic force may add something to a situation but doesn't take away the gravitational force. Likewise, a miracle doesn't violate or contradict the natural order, it supersedes it—and the one who is superseding it is the same one who made the natural order in the first place. So *in a sense* we can take a miracle like another force that is temporarily overruling the physical laws, but unlike one physical law overruling another, it isn't a mindless rule or force: it's an intentional agent directly acting in a way that overrules the laws that he created himself. The laws of nature still operate "all other things being equal," but in the case of a miracle all other things are emphatically not equal: another stronger (and intentional) force is in operation.[4]

The Law of Conservation of Energy says that energy can neither be created nor destroyed—*in an isolated system*. The claim of a miracle is that the universe as a whole is not an isolated system: there is something independent of it that can insert or remove energy from that system at will. So the "all other things being equal" clause that attaches to the Law of Conservation of Energy—and all laws of nature—is that they apply "under natural processes." If a supernatural force decides to act, then these laws simply do not apply.

In Judaism and Christianity, miracles are wonders and signs (John 2:11; 4:54; Acts 2:22; Heb 2:4; etc.). The wonders part is pretty easy to understand: miracles are *wow* moments when things don't go as they usually do. That's true of plenty of natural events too, but in this context they are special acts of God that point to him—they are God's wonders. That leads to the next definition of miracles as signs. This means they are indicative of something, specifically something having to do with God. They are not meaningless interruptions of the way nature usually operates, they *point* to something, most basically that there is a larger reality than the natural order. The miracle makes sense within that larger reality—not implying that it doesn't make sense in our truncated reality, just that we can't fully comprehend it (which shouldn't be too surprising). In plenty of

---

3. C. S. Lewis, *Miracles*, 63–64. Of course, it's much more complicated than this. See McGrew, "Miracles."

4. C. S. Lewis, *Miracles*, 61–62.

miracle accounts, people can see that a miracle is a sign before they know precisely what it's a sign of, besides pointing to this larger reality.[5]

One challenge I hear way too often is, *how* does God perform a miracle? The answer is simple: by willing it. God is alleged to be an all-powerful being, so if he wills something to take place, it takes place. The challenger then asks, what process does God use to push around the matter and energy? The answer is he doesn't use any process. He wills it. The challenge then becomes what *mechanism* does God use to perform these special acts. The answer is, he doesn't use a mechanism. He wills it. Mechanisms operate automatically, without thought, and the claim of a miracle is that an omnipotent, free agent is choosing to perform the act *directly*. There's no intermediary step, no method, no process, no mechanism.

Sometimes the follow-up is, what would you *see* if you observed all this matter being shuffled around without any visible causes? I don't know why I'd be obligated to know the answer to that question, but for now I'll just say: you'd probably see matter being shuffled around without any visible causes. Of course, that's true for a lot of natural events as well (I don't *see* the wind blowing the leaves, I see only the leaves moving), so I don't think it would be a big deal. The difference is that, unlike miracles, the causes of the natural events can be theoretically detectable by further scientific exploration. That's not an insignificant point, but I don't see how it amounts to anything like an objection to miracles.

III. ASSESSING MIRACLES

The claim that a miracle has taken place is the claim that nature is not a closed system, there are other forces in play. Something from the "outside" is doing something within the system so that the physical laws are being overruled in some way; just as if something from outside a system is inserting energy into it, the Law of Conservation of Energy is being overruled for that system. Saying that the physical laws by themselves cannot plausibly account for the event is precisely the point. So with that, there are several questions to ask. First, can miracles happen? If they can, *do* they happen? If they do, can we *know* they have happened?

---

5. C. S. Lewis, *Miracles*, 51, 64–66.

Jesus' resurrection as a test case

The go-to example of a miracle in the philosophical literature is the resurrection of Jesus, presumably because of how central it is to Christianity. I'll just give a quick outline of the idea here. Jesus of Nazareth allegedly did and said things that, in Judaism, only God was allowed to do and say (like forgiving sins). This tended to irk the religious and political establishments of the time; the religious for obvious reasons and the political because by putting himself in God's place, he was portraying himself as the *real* king in opposition to the duly elected king and tyrant. So the religious leaders turned Jesus over to the political leaders, who promptly tortured him brutally and then executed him by crucifixion. A Jesus fan (the term "Christian" had not yet been coined) interred his corpse in a tomb, carved out of solid rock, and then sealed the entrance with a huge stone, which was the common practice. On the following Sunday, a group of his followers found the stone moved and the tomb empty. Then people, some of whom had been opposed to him, began seeing him, talking with him, eating with him, sometimes individually and sometimes in groups numbering in the hundreds. This led his followers to believe that God had physically raised Jesus from the dead—well, that and him telling everyone that's exactly what had happened.

Now if we grant these points for argument's sake, the question becomes whether positing Jesus' actual resurrection would be an acceptable explanation of them.

Hume's objection

David Hume presented the problem of induction.[6] Induction is reasoning from particulars to the general—for example, every swan we've encountered is white, so we inductively infer that *all* swans are white. But valid inductions are still fallible: just because we haven't seen any nonwhite swans doesn't mean there aren't any. Our experiences are not comprehensive and may not be representative. The swan example is a good one because Europeans actually made the induction that all swans are white and when they eventually got to Australia, they discovered black swans.[7] Some inductive inferences are obviously right, some are

---

6. Henderson, "Problem of Induction."

7. Juvenal (first to second century AD), in his sixth satire, said a good wife is like a black swan, i.e., nonexistent. Or Australian. Whichever.

obviously wrong, and many are somewhere in between. Hume argued, however, that we can *never* move from particulars to general. Just because it hurt the first one million times you stuck your tongue in the toaster doesn't give you any reason to think it's going to hurt the million-and-first time. Of course, Hume didn't live his life this way, he was simply arguing that we can't justify inductive inference.[8]

I bring up Hume's critique of induction because Hume goes on to give an inductive argument against miracles that totally contradicts it. He is the last person in the world who should ever have made this objection to miracles.[9] It's weird. I mean, if it was a *good* argument I'd at least have some sympathy; but no such luck.[10]

Basically, Hume argues that the more often we experience something, the more probable it is that it will happen again (like the more often you hurt your tongue by sticking it in the toaster, the more probable it is that it'll hurt the *next* time you stick it in the toaster). We can also aggregate our experiences to see what we experience more often collectively. Those events we human beings have experienced more frequently are more probable to happen again, and those we have experienced less frequently are less probable. When it comes to the regular course of nature, however, we don't simply experience it more frequently; we experience it without exception. That means that the probability of the regular course of nature being interrupted by miracles is not just low, it's zero. We have no experience of such a phenomenon. Therefore, *any* explanation of a unique or anomalous event would be preferable to a miraculous one, regardless of the particular details.[11]

Remember that Sherlock Holmes motto, "When you have eliminated the impossible, whatever remains, however improbable, must be the truth."[12] That's basically what Hume is arguing: miracles are *impossible* so *any* other explanation is better, regardless of how improbable it is. Above, I went over some claims surrounding Jesus' resurrection: 1) Jesus was executed by crucifixion; 2) his corpse was interred in a solid rock tomb; 3) the tomb was found empty a few days later; 4) individuals and groups of hundreds of people interacted with Jesus after his death; 5) the earliest Christian belief was that Jesus was physically, bodily raised from

8. Hume, *Enquiry Concerning Human Understanding*, sec. 4, pp. 11–19.
9. C. S. Lewis, *Miracles*, 106–7.
10. Earman, *Hume's Abject Failure*.
11. Hume, *Enquiry Concerning Human Understanding*, sec. 10, pp. 55–68.
12. Doyle, *Sign of Four*, 93.

the dead. If we grant these for the sake of argument, Hume would argue that Jesus' resurrection is not an acceptable explanation of them since that would have been a miracle, making it maximally improbable. So any other explanation of these claims is preferable.

One alternative explanation—and you're going to think I'm making this up, but I promise it's real—is that Jesus had an evil twin.[13] Seriously. A pair of identical twins were born in Bethlehem the same time Mary gave birth, and one of those twins was accidentally switched with Mary's baby. The switched twin is Jesus. Meanwhile, the other twin grows up somewhere else, but every now and then someone comes up to him and says, "Hey! Jesus!" to which he responds, "Uh, no, my name is Bruce," or whatever. He keeps hearing about this other guy who looks exactly like him, and so finally heads to Jerusalem to find him, and arrives right after Jesus was crucified. Then the twin decides it would be a great idea to impersonate someone who had just been brutally executed for sedition. As ridiculously implausible and improbable and contrived as all this is, it's not impossible and so, by Hume's criteria, it would be a better explanation of those five alleged claims than that Jesus truly rose from the dead.

That by itself should be enough to show that Hume's argument doesn't work, but I can go into more detail. Hume makes some assumptions that are at least controversial if not unjustified. First, he assumes the universe is deterministic. The free will vs. determinism debate is a perennial one, since it's perfectly obvious that both sides are a) correct and b) mutually exclusive. I have a book that's simply a three-hundred(plus)-page bibliography of the free will/determinism issue.[14] Most philosophers, including Hume, embrace soft determinism or compatibilism, which tries to show that we can have free will and determinism, but it's not easy to defend. Hume accepted determinism because he was living in the age of Newtonianism when everyone (other than Newton) thought that you could explain everything by deterministic physical processes.[15] But since then we have discovered the glorious field of quantum physics, which is often understood as fundamentally nondeterministic, although not always.[16]

13. Cavin, "Miracles, Probability, and the Resurrection of Jesus."

14. Rescher, *Free Will*.

15. William B. Ashworth, "Christianity and the Mechanistic Universe," in Lindberg and Numbers, *When Science and Christianity Meet*, 61–84 (80–84 for Newton specifically); Edward B. Davis, "Myth 13," in Numbers, *Galileo Goes to Jail*, 115–22 (esp. 120–21).

16. One interpretation of quantum physics, Bohmian mechanics, is usually thought to be completely deterministic, but see Bohm and Hiley, *Undivided Universe*, 324.

In this case, indeterminism is built into the very structure of the universe, giving us scientific reasons to reject determinism.[17] Supposedly. Neither quantum physics nor the possibility of free will have anything to do with miracles per se, but they do have something to do with Hume's *argument* against miracles, since that argument is dependent on the universe being absolutely determined. Insofar as determinism is questionable, Hume's argument is questionable.

Second, Hume assumes the universe is *static*, that the more often something happens in the past, the more often it will happen in the future. But even a deterministic universe need not be static, there is still plenty of room for rare and unique events to occur. Moreover, this would apply to rare and unique *natural* events just as much as to miracles. Meteor strikes are rare; rogue waves are rare; but that doesn't mean they can't be good explanations—even better explanations than those that appeal to more frequent events. As of July 19, 1969, no one had ever walked on the moon, but that didn't make it maximally improbable that two guys would walk on it the next day. But Hume's assumption of a static universe would apply here and to many other natural events, making them maximally improbable. Since they're not maximally improbable, the assumption of a static universe is unjustified.

Just to belabor the point, let me give an unnecessarily complicated thought experiment. Say there's a naturally elevated basin that feeds a river running through a forest. Precipitation provides just a *bit* more water into the basin than it puts out through feeding the river and evaporation. Over a million years it builds up, and the basin walls erode, until they cannot contain the weight of the water, and collapse, spilling a huge tsunami over the forest and destroying it. This could be the first and only time such an event happens, yet it could have happened in a deterministic universe; it could even have been predictable. Again, this doesn't address the possibility of miracles, it addresses only Hume's argument against miracles. For his argument to work, the frequency of past events must translate into their probability. But then we would have to say events like I just described are maximally improbable and have a 0 percent chance of occurring since they would never have happened before. Obviously that isn't true.

Hume's third assumption is the big one: the universe is a closed system. Remember, physical laws are defined by how they operate in

---

17. Hilgevoord and Uffink, "Uncertainty Principle."

closed systems, and they have "all other things being equal" rider clauses. When we're discussing miracles, the question is whether the physical universe as a whole is a closed system. Since physical laws describe only how events normally transpire, their rider clauses say they apply "under natural conditions." But the very claim of a miracle is that it's *not* under natural conditions; something from outside nature is producing an effect within it. Sure, you can reject this, but you can't reject it *and then use this rejection as your reason for rejecting it*. Hume assumes that the universe is a closed system to argue that the universe is a closed system. This is blatantly circular.[18] He starts by saying we have universal experience against miracles, but for this to be the case miracles would never have happened. If miracles *have* happened, then we don't have universal experience against them. Basically, he's arguing

1. The universe is a closed system,
    therefore,
2. Miracles never happen,
    therefore,
3. We have universal experience against miracles,
    therefore,
4. Miracles are maximally improbable,
    therefore,
5. Miracles never happen,
    therefore,
6. The universe is a closed system.

IV. EXPLANATORY VIRTUES

Are miracles improbable?

OK, ignoring all that, let me just address the fourth premise that miracles are maximally improbable, or at least extremely improbable. They are certainly extremely rare or unique—that's precisely why the Bible includes accounts of them: they are noteworthy, out-of-the-ordinary events (Deut 4:32–35). In the Reformation, Protestant Christians argued that miracles and revelation go together. As such, we shouldn't expect miracles to

---

18. C. S. Lewis, *Miracles*, 105–6.

happen today, "for miracles are ceas'd."[19] Today, plenty of Protestants disagree with this, as do Catholic and Orthodox Christians.[20]

At any rate, miracles, as commonly understood, are extremely rare or even unique. But does this translate into a low probability? Because if it does, this would apply to rare and unique *natural* events too, like meteor strikes, rogue waves, and the first moon landing. There are two ways we can look at this, based on whether we're using probability in a predictive sense or an explanatory sense. We usually can't predict miracles or meteor strikes: at any given moment, the likelihood that there will be a meteor strike right where you happen to be standing is incredibly low. No moment and no location is more privileged than any others in this regard. But that's because we don't have specific information to suggest it. Saying we don't have it, however, doesn't mean the information isn't out there, just that it's not available to us—information like the specific location, trajectory, and impetus of a particular meteor as it relates to the earth and its motion. If we had access to that information, we *could* predict it.

But of course, the fact that we don't have information allowing us to predict it doesn't mean we can't use a meteor strike as an explanation after the fact, because we now have other information: there's a crater in the ground; there's a big swath of forest that's been quickly burned out; perhaps there's a big rock in the crater that is composed of metals not local to that region, producing colors never before seen on our earth.[21] If, in the face of this evidence, someone were to object that the probability *beforehand* that a meteor would strike at that particular location at that particular time was insanely low we'd think he was an idiot. We have very good reasons for thinking a meteor strike took place at that location at that time, making the probability very high. Saying it's low because we couldn't have predicted it beforehand is inane.

Now let's move from meteors to miracles. We generally can't predict miracles any more than we can predict meteor strikes. But this isn't because miracles are random; they are *signs*, events caused directly by God. Is there information that we don't have access to that would allow us to predict a miracle? *Of course*: God's intentions. If we knew that God was intending to produce a particular effect at a particular place at a particular time, we could predict that it would happen. The probability would be

---

19. Shakespeare, *Henry V*, act 1, scene 1.

20. Warfield, *Counterfeit Miracles*.

21. If you get this reference, you'll also appreciate that I have a Cthulhu fish on the back of my car.

one. Can we appeal to miracles after the fact to explain certain phenomena? *Of course.* For example—and I'm just pulling this out of the air—if the claim is that a man who was dead for a few days was resurrected, then we would expect phenomena such as that his tomb would be empty; that people saw him alive, then dead, then alive again; that kind of thing. If these data points were given, then we could potentially offer a miraculous explanation of them, in the same way that we can offer a meteor strike as an explanation of the crater.

Naturally there's a lot more to say here to muddy the waters. We may not know the location and trajectory of a meteor before it strikes, but the way we could have access to this information is through science. Scientists often predict near misses by small space detritus, and it's only a matter of time before they'll accurately predict that a space chunk[22] will strike the earth at a particular place and time. The information that would allow us to predict a miracle is not available to science because it is not mechanistic nor random nor natural but the free intentions of a supernatural, intelligent agent. We could predict it only if God let us know what his specific intentions were beforehand. Without access to this information, we cannot make any kind of *predictive* probability assessment. So meteor strikes and other natural phenomena are predictable in a way that miracles are not. It's up to *us* to *discover* whatever evidence would allow us to predict a natural event, but it's up to *God* to *reveal* whatever evidence would allow us to predict a miracle.

History is like science in that it looks for the evidence that—had it been known at the time—could have been used to predict the result. However, history involves the study of people making free choices that are neither mechanistic nor random. So, given free will (which usually isn't given, by the way), history would not allow for as much predictability as science does. Even so, history seeks to find the rational motives behind free choices, assuming that they are discoverable, thus making their choices somewhat predictable. So science and history take probability in both predictive and explanatory senses, since both involve evidence that is theoretically available to us to discover. Miracles can take probability in an explanatory sense since we can potentially discover the evidence of a supposed miracle (like an empty tomb or postmortem appearances). But miracles can't take probability in a predictive sense, since the evidence

---

22. This was my nickname in high school.

that would allow us to predict one is not up to us to discover but up to God to reveal.

So perhaps if we take "historical explanation" in a technical sense where it may require both senses of probability, we can't appeal to miracles as historical explanations. But if we take it in a nontechnical sense, where it just means "a rational explanation of historical data," then being limited to the explanatory sense of probability without the predictive sense is not a problem.

Of course, if God doesn't exist, then the probability of God performing a miracle is zero. So the probability of miracles in general (not particular instances of them) corresponds pretty closely to the probability of God's existence. If God *might* exist then miracles *might* occur, and we can't rule them out of court a priori. Therefore, to say miracles can't occur, we would have to first demonstrate that God can't exist. Unless we know God doesn't exist, miracles are possible and not inherently improbable.

Are miracles ad hoc?

One last objection I'll deal with is that, if we're allowed to explain something as a miracle, then we can explain anything and everything as a miracle. "How did this unusual thing happen?" "Goddidit!" (all one word). "Why does the sun rise in the east?" "Goddidit!" And so on. Basically, this objection is that if we allow miracles into the fold, we can appeal to them in a contrived or ad hoc way; therefore, *any* appeal to a miracle is contrived or ad hoc. But this goes against the very concept of miracle as "sign." A miracle is supposed to be indicative of something, it is not a random interruption of the natural order. It says that there is a larger reality and the miracle makes sense within that reality. Naturalists "have mistaken a partial system within reality, namely nature, for the whole."[23] And saying it makes sense within the larger reality doesn't mean it makes *no* sense in our condensed reality; we can't fully comprehend its meaning but we can still see that it is a sign of something, it is not random or ad hoc. We can see that a miracle has a purpose before we can see what that purpose is. If miracles are supernatural acts of God, they would mean something even if we don't know precisely what.[24]

---

23. C. S. Lewis, *Miracles*, 64.

24. C. S. Lewis, *Miracles*, 64–66. Because of my particular neuroses, I've tried to find a loophole to this idea: What if God performed miracles that made no sense to indicate (as a sign) that things *are* random? But even if they were signs that things are random,

Now, certainly miracles can be appealed to in an ad hoc way, and often are. But so can any other concept, any other kind of explanation. Some Marxists use economics in an ad hoc way, saying all societal change is the product of economic factors (Economicsdidit!). But that wouldn't justify denying that there are economic forces, that they sometimes (or often) produce societal changes, or that economics is a valid field of study. Just because *some* appeals to economics as explanations are ad hoc doesn't mean that *all* such appeals are.

Ad hoc explanations don't fit well with the data to be explained. We tend to have an intuitive sense of this, but when it gets into gray area, people disagree. Roughly, the more you have to go beyond the data to be explained, the more ad hoc an explanation is. Of course, *any* explanation goes beyond the data just by virtue of not being a piece of data itself, but the further beyond you have to go, the more ad hoc it is. However, it's one of those aspects of an explanation (including scientific explanations) that goes more by feel, like parsimony or beauty, which often overlap ad hocness quite a bit.[25] Ad hocness is the meat and potatoes of conspiracy theories. They take some of the evidence that is difficult to fit into the accepted explanation and form a new explanation based on them while ignoring the vast majority of evidence that fits perfectly into the accepted explanation and doesn't fit into the conspiracy theory.[26]

Back to miracles. The problem with appealing to miracles, supposedly, is that they appeal to a "god of the gaps." The claim here is that there are gaps in our knowledge—we're not sure how things go from point A to point B—so we plug in that Goddidit! formula to bridge the gap. Then science comes along and removes the gap by showing how natural processes explain how to get from A to B. I've already addressed this to some extent:[27] some phenomena were thought to be too recalcitrant to be explained by natural processes and so were thought to be direct acts of God, but later evidence showed natural processes *could* explain them (think evolution). But there are also other phenomena that were thought to be easily explicable by natural processes that later evidence showed

---

the miracles themselves would not be random by virtue of being deliberate attempts to communicate a specific message. It's kind of like Diogenes in ancient Greece who would pee, poop, and do coarser things in public to show that human beings are mere animals. But mere animals don't pee and poop in public to *make a point*.

25. "Sometimes science is more art than science, Morty. A lot of people don't get that."—Rick Sanchez

26. Schindler, *Theoretical Virtues in Science*, 119–20.

27. See above, 135–36.

were agonizingly recalcitrant (think origin of life). This kind of objection ignores one half of the issue and portrays the other half as if it represents the whole. Moreover, any explanation could be used in this way: e.g., our Marxists would be guilty of appealing to an economics of the gaps. Some people are guilty of a naturalism of the gaps: they gloss over gaps in our information by appealing to natural processes that simply don't fit or have not been discovered.[28]

Recall the allegations surrounding Jesus' resurrection: 1) Jesus was killed by crucifixion; 2) his corpse was placed in a solid rock tomb; 3) this tomb was found empty a few days later; 4) people saw, spoke with, and physically interacted with Jesus after his death; and 5) people believed Jesus had literally, physically risen from the dead immediately after it supposedly happened. If we grant these data points for the sake of argument, explaining them with the miracle that Jesus rose from the dead fits the data, it naturally emerges from them. The only extra datum we would have to add is that a God capable of doing it exists—and for most people that wouldn't be an extra datum since most people already believe in God.

Contrast this with the evil twin theory: we have no reason from the data alone to suggest such a scenario, and we have to keep adding details to it not given by the mere presence of a twin to explain the data. Identical twins are born at the same time and place as the baby Mary has given birth to; one of these twins is switched with Mary's baby; the other twin is raised geographically distant from the switched twin (Jesus); he comes to Jerusalem right after the crucifixion; he decides to continue Jesus' ministry under false pretenses; he steals his corpse from the tomb and impersonates him; and then he disappears without a trace. Those steps do not follow from what precedes them, and there's no reason to accept any of them, they're just pulled out of the air. The evil twin theory is ridiculously ad hoc, but the resurrection theory (if you want to call it a theory) is not. This is important: the naturalistic explanation is ad hoc and the miraculous explanation is not. The same goes for those who suggest that Jesus was an extraterrestrial or a time traveler (yes, there are people who say they believe this). They prefer a naturalistic explanation that is absurdly ad hoc over a supernaturalistic explanation that is not ad hoc. These explanations are guilty of appealing to a naturalism of the gaps.

---

28. Nagel describes a "Darwinism of the gaps," where people use evolutionary ideas to cover up our gaps in knowledge (*Mind and Cosmos*, 127). Cf. below, 250n41. Maybe we could call it "Gaia of the gaps."

Gaps arguments are bad explanations precisely because they are ad hoc. And of course, plenty of people *do* use the possibility of miracles to explain things in an ad hoc way. But it doesn't follow that all claims of miracles are contrived and ad hoc or guilty of appealing to a god of the gaps. "Whatever men may *say*, no one really thinks that the Christian doctrine of the Resurrection is exactly on the same level with some pious tittle-tattle about how Mother Egarée Louise miraculously found her second best thimble by the aid of St. Anthony."[29] The reason gaps arguments are ad hoc is that they base themselves on what is unknown rather than on what is known. Applying it to the resurrection, the claim would be that, just because we have not yet discovered natural processes that could explain how a man dead for three days could be resurrected with a glorified body, it doesn't mean that forthcoming evidence will not supply a naturalistic explanation. But this is a blatant appeal to a naturalism of the gaps. When Christians claim that God raised Jesus from the dead, it is not based on what is unknown but on the alleged data. Of course, the data may be demonstrated to be inaccurate or the explanation may fail for other reasons—but that doesn't make it a gaps-type of argument.

If we have *specific reasons* for thinking a miracle took place, then it is not ad hoc or an appeal to a god of the gaps to suggest it as a possible explanation, whatever other failings it might have. In the same way, if we have specific reasons for thinking a meteor strike or a rogue wave took place, it is not ad hoc to suggest it just because we couldn't predict it beforehand. The specific reasons for thinking a meteor strike took place would be the crater, the detritus inside it, the destroyed landscape around it, etc. The specific reasons for suggesting that God raised Jesus from the dead are those five claims surrounding it, if they can be shown to be correct, and the context in which it happened. Jesus was tortured and executed by crucifixion for blasphemy (putting himself in God's place) and treason (claiming to be a king). As Pannenberg writes, the significance of Jesus' resurrection is not that some guy came back to life, it's that *this particular guy* did. For God to raise *that* guy amounts to his divine vindication. It's a sign that he was and is who he claimed to be and had the authority he claimed to have.[30] Again, it could still be false, it could still fail to be a good explanation for a number of reasons. But not for being ad hoc. Not for appealing to a god of the gaps. It isn't and doesn't.

---

29. C. S. Lewis, *Miracles*, 111.
30. Pannenberg, *Glaube und Wirklichkeit*, 92–94.

POSTSCRIPT I: THOUGHTS AND PRAYERS

There are two tangential issues that relate to miracles and science that I want to briefly discuss. The first is the efficacy of prayer: if God answers prayers, then it would mean he changes the way the course of events were proceeding to produce a different effect that would not have taken place without the prayer. Does prayer tend to be answered in the sense of regularly and predictably producing the results requested? The Bible already answers this in the negative: Jesus prayed, "May this cup be taken from me" (Matt 26:39), the cup in question being his forthcoming crucifixion; said cup *didn't* pass from him, and he was crucified and killed. So Jesus' prayer did not produce the requested results.

The reason for this should be obvious: God is not a cosmic gumball machine. We make *requests* of him, not demands. Physical laws take place with a nomic necessity—that is, given their "all other things being equal" clause, we can depend on their constancy. This is what allows us to know what would have happened when we interfere with the physical processes and objects in question: we can contrast what happens when we're not interfering with what happens when we are. But since God is not a mindless mechanism, we can't make this contrast with prayer.

> The question then arises, "What sort of evidence *would* prove the efficacy of prayer?" The thing we pray for may happen, but how can you ever know it was not going to happen anyway? Even if the thing were indisputably miraculous it would not follow that the miracle had occurred because of your prayers. The answer surely is that a compulsive empirical proof such as we have in the sciences can never be attained. . . . The essence of request, as distinct from compulsion, is that it may or may not be granted. And if an infinitely wise Being listens to the requests of finite and foolish creatures, of course He will sometimes grant and sometimes refuse them. Invariable "success" in prayer would not prove the Christian doctrine at all. It would prove something much more like magic—a power in certain human beings to control, or compel, the course of nature.[31]

There have been suggestions, however, to try to test scientifically the efficacy of prayer.[32] If we could get a group of people to pray for the recovery of everyone in one particular hospital, then we could contrast

---

31. "The Efficacy of Prayer," in C. S. Lewis, *World's Last Night*, 4–5.
32. Robert B. Mullin, "Science, Miracles, and the Prayer-Gauge Debate," in Lindberg and Numbers, *When Science and Christianity Meet*, 203–24.

the rate of recovery of the patients there with those of a control hospital that no one prayed for. Or perhaps we can have a group of one religion or denomination pray for one hospital and a group of another religion or denomination pray for another and see who comes out on top.

But prayer involves confession, penitence, adoration, and other elements, it's not just asking God to do something. The latter is called petitionary prayer. And doing it to *test its efficacy* is not consonant with theistic religions.

> The trouble is that I do not see how any real prayer could go on under such conditions. "Words without thoughts never to heaven go," says the King in *Hamlet*. Simply to say prayers is not to pray; otherwise a team of properly trained parrots would serve as well as men for our experiments. You cannot pray for the recovery of the sick unless the end you have in view is their recovery. But you can have no motive for desiring the recovery of all the patients in one hospital and none of those in another. You are not doing it in order that suffering should be relieved; you are doing it to find out what happens. The real purpose and the nominal purpose of your prayers are at variance. In other words, whatever your tongue and teeth and knees may do, you are not praying. The experiment demands an impossibility.[33]

Obviously there's a lot more—like a *lot* more—to be said about petitionary prayer. I'm only dealing with it here as it relates to science and religion, as a tangent of the problem of miracles. There are plenty of objections that have been raised about it that I'm not even mentioning here, so you'll still have to deal with them.

POSTSCRIPT II: NEAR-DEATH EXPERIENCES

There've been a couple of movies called *Flatliners* about medical students who kill themselves in nonviolent ways in order to be revived afterwards, because *that's* not insane. The goal was to see if there's an afterlife, and if there is, what it's like. "Philosophy failed. Religion failed. Now it's up to the physical sciences," says Kiefer Sutherland's character. And there are plenty of accounts of real people who die and say they experienced an afterlife where they meet with their deceased loved ones, or talk with God, or experience the torment of hell, but are then revived.

---

33. "The Efficacy of Prayer," in C. S. Lewis, *World's Last Night*, 6.

One problem I have with this is the same as with the efficacy of prayer: it assumes there is a mechanistic, automatic process involved such that we go to the afterlife as soon as our brain functions cease. But in theistic religions, in addition to being a free agent, not mechanistic, God is also omniscient—that is, he knows all true propositions. So he knows if someone is truly dying or if they're going to be revived. Now, it's certainly possible he set up entry into the afterlife as a mechanistic process so that we go there once we have no brain function whether we're going to stay that way or not, but I don't see any reason to accept this. I'm just skeptical of the idea that we can test God in this, or any other, way.

Another problem I have with near-death experiences is how exactly do they know that the person had the experience during that short period when they had no brain functions? I mean, they're called *near*-death experiences, not *during*-death experiences. We all know that we can have vivid and extended dreams while asleep that barely take up any time. How do we know near-death experiences weren't dreams experienced in a flash right before the patient bottomed out or as soon as they came back on? Perhaps you could say such a flash would involve a lot of neurons firing and so would be indicated on an EEG, but I don't think the absence of such an indication is going to amount to much—not least because I don't think the mind can be reduced to the physical brain.

Bear in mind that there are Christians much smarter than I who take near-death experiences seriously,[34] I'm just skeptical of them for the above reasons. At any rate, I don't think it can be explored scientifically, like they tried to do in the *Flatliner* movies. Of course, I could be wrong. In fact, yes, that seems more likely.

---

34. Habermas and Moreland, *Immortality*, 73–105.

Chapter 14

Theism and the Justification of Science

I WENT OVER THE history of science and religion in chapter 12, but kind of glossed over an interesting point: the Scientific Revolution took place *within Christian civilization*. Christian beliefs, doctrines, practices, and the Bible itself greatly influenced the development of science.[1] There was more to it, of course; they weren't discarding everything and building from the ground up. But if science and Christianity are at odds, isn't it at least *unusual* that its early champions were nearly all Christians, acting from explicitly Christian motives?

I. TWO QUESTIONS

Some people argue from this that Christianity uniquely allows for science and that a Scientific Revolution wouldn't happen under other worldviews. Stanley Jaki, a priest and physicist, wrote, "The fundamental paradigm of science [is] its invariable stillbirths in all ancient cultures and its only viable birth in a Europe which Christian faith in the Creator had helped to form."[2] Sociologist Rodney Stark writes,

---

1. In addition to Hannam's *God's Philosophers*, I recommend some of Peter Harrison's works: *The Bible, Protestantism, and the Rise of Natural Science*; *The Fall of Man and the Foundations of Science*; and "Development of the Concept of Laws of Nature."
2. Jaki, *Road of Science and the Ways to God*, 243.

Theism and the Justification of Science

Had the followers of Jesus remained an obscure Jewish sect, most of you would not have learned to read and the rest of you would be reading from hand-copied scrolls. Without a theology committed to reason, progress, and moral equality, today the entire world would be about where non-European societies were in, say, 1800: A world with many astrologers and alchemists but no scientists. A world of despots, lacking universities, banks, factories, eyeglasses, chimneys, and pianos. A world where most infants do not live to the age of five and many women die in childbirth—a world truly living in "dark ages."[3]

Well . . . maybe. But it's also possible that proto-science had developed to a critical mass and would have spilled over into a Scientific Revolution regardless. Christianity might have just been in the right place at the right time. If Islam had been in the right place at the right time, would we be able to find reasons why the Islamic concept of God uniquely allows for science? If China was in the right place at the right time, would we be able to find reasons why the concept of *tien* (sky or heavens) in Chinese religions uniquely allows for science?

Really, this is not one question but two. The first is *The Philosophical Question*™: Is science dependent on the Judeo-Christian worldview or a close facsimile thereof? Must science ultimately *presuppose* something like Christianity or can we make sense of science and its presuppositions without it? Note that the question is *not* whether individual scientists must assume Christianity is true to do science. The question is whether the preconditions that must be true for science to be valid—like that the world is comprehensible and our minds are capable of understanding it—can be justified apart from a worldview like Christianity. And if so, are other religions and nonreligious worldviews close enough to Christianity to secure or make probable these preconditions, or are Judaism and Christianity the only options available?[4]

The second question is dependent on the first: If the answer to the Philosophical Question is yes, is that *why* the Scientific Revolution took place within a Christian milieu? This is *The Historical Question*™. But this is enormously difficult to answer as there were many other factors besides Christianity involved in the development of science, some of which

---

3. Stark, *For the Glory of God*, 233.

4. Regarding the second point, whether our minds are capable of understanding, and how the various religious traditions in the world make this more probable or less, see Baldwin and McNabb, *Plantingian Religious Epistemology and World Religions*.

are unknown, some of which are other religions, and some of which are not religious in nature (like commerce).⁵ Plus, it assumes that culture, society, and civilization are all rational in that they follow the logical consequences of their beliefs to their conclusions, something I think is not in evidence. Ideas have consequences, certainly, and different ideas have different consequences, but they won't always be *rational* consequences. The best case that can be made for answering yes to the Historical Question, I think, was made by Alfred North Whitehead.

> I do not think, however, that I have even yet brought out the greatest contribution of medievalism to the formation of the scientific movement. I mean the inexpugnable belief that every detailed occurrence can be correlated with its antecedents in a perfectly definite manner, exemplifying general principles. Without this belief the incredible labours of scientists would be without hope. It is this instinctive conviction, vividly poised before the imagination, which is the motive power of research:—that there is a secret, a secret which can be unveiled. How has this conviction been so vividly implanted on the European mind?
>
> When we compare this tone of thought in Europe with the attitude of other civilisations when left to themselves, there seems but one source for its origin. It must come from the medieval insistence on the rationality of God, conceived as with the personal energy of Jehovah and with the rationality of a Greek philosopher. Every detail was supervised and ordered: the search into nature could only result in the vindication of the faith in rationality. Remember that I am not talking of the explicit beliefs of a few individuals. What I mean is the impress on the European mind arising from the unquestioned faith of centuries. By this I mean the instinctive tone of thought and not a mere creed of words.
>
> In Asia, the conceptions of God were of a being who was either too arbitrary or too impersonal for such ideas to have much effect on instinctive habits of mind. Any definite occurrence might be due to the fiat of an irrational despot, or might issue from some impersonal, inscrutable origin of things. There was not the same confidence as in the intelligible rationality of a personal being. I am not arguing that the European trust in the scrutability of nature was logically justified even by its own theology. My only point is to understand how it arose. My explanation is that the faith in the possibility of science, generated

---

5. Cook, *Matters of Exchange*.

antecedently to the development of modern scientific theory, is an unconscious derivative from medieval theology.[6]

So the Scientific Revolution occurred because people believed that "every detailed occurrence can be correlated with its antecedents in a perfectly definite manner, exemplifying general principles"; and they believed *this* because of "the medieval insistence on the rationality of God, conceived as with the personal energy of Jehovah and with the rationality of a Greek philosopher." Other religions, civilizations, and cultures apparently don't have these elements, or at least don't have them as strongly. God may be an "irrational despot," God may be an *immoral* despot, the universe may be unreliable and variable, and the human mind might be untrustworthy. Without belief in "the intelligible rationality of a personal being" we would have no confidence that reality is rationally structured or that our minds are capable of rationally comprehending it, which would prevent the necessary conditions for science from taking hold. So Greek philosophy + Christianity = science.

But there's an obvious objection: If this is the case, why did it take so long? We had Greek philosophy for over two millennia and Christian theology for a millennium and a half before the Scientific Revolution occurred.[7] But Whitehead has a response: "Remember that I am not talking of the explicit beliefs of a few individuals. What I mean is the impress on the European mind arising from the unquestioned faith of centuries. By this I mean the instinctive tone of thought and not a mere creed of words." So let's amend the recipe. Combine Greek philosophy with Christianity in a bowl, whisk until they're nice and fluffy, and then bake at 350° for 1,500 years, until your Scientific Revolution is bubbling over.

And once again my answer is: well . . . maybe. If we grant a positive answer to the Philosophical Question, the above seems moderately plausible, but I still can't make my way around the assumption that culture and society must be rational for this to work. Plus, there are so many *other* factors that would have to be in place for a Scientific Revolution to occur that have little to nothing to do with whatever worldview a culture

---

6. Whitehead, *Science and the Modern World*, 13–14. Note that Whitehead gives us the only scenario allowing us to answer yes to the Historical Question and no to the Philosophical Question: "I am not arguing that the European trust in the scrutability of nature was logically justified even by its own theology." That is, they may have *mistakenly* thought that Christian theology taught or entailed these things.

7. Not to mention that Judaism could supply many of the elements that Christianity does, and it's been around for much longer than Greek philosophy.

accepts. I don't mean to be overly cynical about this, but the problem with studying history is that you're searching for rational explanations for why irrational people did irrational things in irrational cultures.

Regarding the issue at hand, many people don't distinguish the Philosophical from the Historical Question. Any argument in favor of the former is often taken as an argument for the latter, and any argument against the latter is taken as an argument against the former. As you can probably tell, I'm suspicious of the Historical Question. But that's OK because I'm a philosopher, so I'm going to spend the rest of this chapter talking about the Philosophical Question. Aren't you lucky!

II. THE EUTHYPHRO DILEMMA

So is there some sense in which Christianity uniquely allows for science? One way to address this comes from the Euthyphro dilemma, which comes from one of Plato's Socratic dialogues.[8] The question Socrates and Euthyphro ask is whether God commands the good because it is good or is the good good because God commands it? But neither way is acceptable. If God commands the good because it is good, it means that the good is a higher reality than God, and God commands us to be good because he must align himself with this higher reality just as much as we do. But then God wouldn't be the ultimate reality, the supreme being, and so he wouldn't be God—if we are going to worship something, we should worship the higher reality. On the other hand, if the good is good because God commands it, then it is arbitrary. If "the good" simply means "what God commands," then if God commanded hatred, murder, and rape, these would be the good.

Now one thing that drives me freaking nuts is when philosophers, professional philosophers, bring up the Euthyphro dilemma as if no one has offered any rebuttals to it over the last few millennia. I mean, if they went over the rebuttals and showed that they fail, that would be one thing. But they don't. They usually don't even mention that there *are* rebuttals. Of course, it's open to grasp either horn of the dilemma. Some theists—specifically Muslims, but also some Christians—accept that the good is good because God commands it. But the traditional response in Christianity is to say that it's a false dilemma. It puts the two ideas into a dependence relationship with each other: if God's command explains

---

8. Plato, *Euthyphro*.

why actions are good, then they are arbitrary; but if actions' goodness explains why God commands them, then they are a higher reality than he is. But what if they're *not* in a dependence relationship with each other?

The standard Christian rejoinder through the ages is that these two concepts are both aspects of a *third* thing: God's nature. The good isn't arbitrary, it is intrinsically good and it could not be otherwise. But its goodness is not something outside of God, it is an expression of God's own nature, his intrinsic goodness. It's not a higher reality than him, it *is* him. God is the ground of morality. By this I mean he is not merely the *source* of morality, since this is compatible with him arbitrarily deciding what the good is. He is the ground of morality as well as the ground of reality. This provides an objective anchor for morality, tying it to ultimate reality itself. It's not enough for the ultimate good to be a precept; it's a person.[9]

III. GROUNDING RATIONALITY

"OK, fine, but what does this have to do with whether science must presuppose something like Christianity?" I hear you cry in a kind of whiny voice. Just this: what we can say about the principles of morality also applies *mutatis mutandis*[10] to the principles of rationality. Just as morality is an expression of God's nature, so rationality is an expression of God's nature. God is the ground of rationality, morality, and reality. This is why Jesus of Nazareth is called the *logos*, the incarnated cosmic principle of rationality (John 1:1–18). And when we recall the doctrine of *imago Dei*, that we are created in God's image, this means we are (occasionally) rational beings, and by being rational we are participating in the divine nature. We are rational agents because we are created by and in the image of the ground of rationality.

Moreover, if the ground of rationality is also the ground of reality and the creator of the physical universe, then we should expect the physical universe to be rational—not in the same sense that *minds* are rational but in the sense of being comprehensible by rational minds. We should expect the universe to make sense, at least on some level. The claim being made

---

9. For historical responses along these lines, see Burt, "Augustine and Divine Voluntarism"; Rogers, "St. Anselm of Canterbury on God and Morality"; Koritansky, "Thomas Aquinas and the Euthyphro Dilemma." Or you could, you know, actually read Augustine, Anselm, and Aquinas. For contemporary responses, see Robert Adams, *Finite and Infinite Goods*, 13–49, and Alston, "What Euthyphro Should Have Said."

10. Latin for "Shut up and agree with me."

is not that someone has to consciously believe in God in order to believe that we are rational and that the universe is understandable, any more than someone has to consciously believe in God in order to recognize that murder is immoral. We know these things intuitively. The suggestion is that Christianity provides a metaphysical backing that gives us a reason for *trusting* those intuitions beyond the mere fact of having them.

Having said this, the idea that God is the ground of rationality is more complex than I've portrayed it. Plantinga points this out by contrasting Aquinas with William of Ockham.

> What does this "rationality" of God consist in? What might the medieval have meant in saying that God is rational? When they discussed this topic, the medieval put it in terms of the question whether, in God, it is *intellect* or *will* that is primary. They thought that if *intellect* is primary in God, then God's actions will be predictable, orderly, conforming to a plan—a plan we can partially fathom. On the other hand, if it is *will* that is primary in God, then his actions would involve much more by way of caprice and arbitrary choice and much less predictability. If it is intellect that is prior in God, then his actions will be rational—rational in something like the way that we are rational; if it is will that is prior, then one can't expect as much by way of rationality. Aquinas championed the primacy of intellect in God, while William of Ockham endorsed the priority of will. This, of course, is vastly oversimplified (as is nearly anything one can say about medieval philosophy) but it conveys an essential point. Ockham seems to think that God's will was essentially unconstrained by God's intellect (or anything else); God is free to do whatever he wants, even something that is irrational in the sense of contrary to what his intellect perceives as good or right. . . . He also claimed that God could have commanded hatred instead of love, adultery instead of faithfulness, cruelty instead of kindness; and if he had then those things would have been morally obligatory. Aquinas, on the other hand, taught that God's commands stem from his very nature, so that it isn't so much as possible that God should have commanded hate rather than love.
>
> The rationality of God, as Aquinas thought, extends far beyond the realm of morality. God sets forth moral laws, to be sure, but he also sets forth or promulgates laws of nature, and he creates the world in such a way that it conforms to these laws. The tendency of Ockham's thought, on the other hand, is to emphasize the freedom (willfulness?) of God to such a degree that he becomes completely unpredictable; and to the extent that

God is completely unpredictable, the same goes for his world. There is no guarantee that the world at some deep level is law governed, or lawful; there is no guarantee that God's world is such that by rational, intellectual activity, we will be able to learn something about its deep structure. In fact there is no reason to think, on Ockham's view, that it *has* a deep structure.... Modern science required a sort of instinctive conviction that God is more like the way Aquinas thinks of him than the way Ockham does. And indeed, many of the early pioneers and heroes of modern western science, the scientists propelling the Scientific Revolution, clearly sided with Aquinas.[11]

Historically, Aquinas's view, or something like it, has been the majority position in Christianity, and Ockham's view, or something like it, has been a minority one (this is not to say that those who agree with Ockham are foolish). Also, note how closely this connects with the issue of morality: if God could do something that his intellect perceives as irrational, he could also do something his intellect perceives as immoral. Practical reason is the use of reason to guide our actions, and this overlaps ethics quite a bit.

So Ockhamistic Christianity doesn't provide support for science and possibly undermines it, but we can comfort ourselves in the knowledge that it hasn't been as influential as Thomistic Christianity has been. But what about other religions? Apart from Judaism, the closest religion to Christianity is Islam. Does it say God's nature is such that his intellect is prior to his will? Well, there's a double whammy here: first, no, *nothing* is prior to God's will in Islam, because if something were, it would limit his omnipotence and there can be no limit to what God can do. Second, in Islam, God doesn't even have a nature, for pretty much the same reason: if God had a nature, then it would mean he could not act contrary to it (otherwise it wouldn't be his *nature*) and so this would, again, limit the expression of his power.

In 2006, Pope Benedict XVI quoted a medieval debate on this point in his Regensburg lecture and got a lot of heat for it. The debate he was quoting, translated by Theodore Khoury, was between "the erudite Byzantine emperor Manuel II Paleologus and an educated Persian on the subject of Christianity and Islam, and the truth of both."[12] The emperor argued that, in Christianity, God has a nature and a principal aspect of

---

11. Plantinga, *Where the Conflict Really Lies*, 272–74; emphasis in original.
12. Benedict XVI, "Faith, Reason and the University."

this is intellect, rationality, and reason (both pure and practical). In Islam, however, God does not have a nature since this would restrict his omnipotence: God must be free to do anything, including being irrational and immoral. In the dialogue, the emperor explains

> why spreading the faith through violence is something unreasonable. Violence is incompatible with the nature of God and the nature of the soul. "God," he says, "is not pleased by blood—and not acting reasonably (σὺν λόγω) is contrary to God's nature. Faith is born of the soul, not the body. Whoever would lead someone to faith needs the ability to speak well and to reason properly, without violence and threats.... To convince a reasonable soul, one does not need a strong arm, or weapons of any kind, or any other means of threatening a person with death...."
>
> The decisive statement in this argument against violent conversion is this: not to act in accordance with reason is contrary to God's nature.... But for Muslim teaching, God is absolutely transcendent. His will is not bound up with any of our categories, even that of rationality. Here Khoury quotes a work of the noted French Islamist R. Arnaldez, who points out that Ibn Hazm went so far as to state that God is not bound even by his own word, and that nothing would oblige him to reveal the truth to us. Were it God's will, we would even have to practise idolatry.[13]

So, by this account at least, the Islamic concept of God is not only unconstrained by rationality, but unconstrained by morality as well. When looking back at the Euthyphro dilemma, Islam portrays God in one of the binary patterns: the good is good because God commands it, so that the appellation "good" just means "whatever God commands"; if he commanded something else, then *that* would be good. In the same way, the rational is rational because God dictates it, so that whatever God does is rational, and if he chooses to do the opposite, then *that* is the rational. The God of Islam is not the ground of morality and rationality, he is the source or creator of them.

By contrast, as we've seen, Christianity says morality and rationality are not arbitrary decisions God makes, they are expressions of his very nature, the nature of a metaphysically necessary being.[14] He doesn't create

---

13. Benedict XVI, "Faith, Reason and the University."

14. God would be metaphysically necessary *if* he exists since a supreme being could hardly be supreme if he/she/it/they were contingent. But if he doesn't exist then obviously he wouldn't be metaphysically necessary since he wouldn't even be metaphysically *actual*.

them, choosing what they should be, he grounds them. The God of Islam is completely transcendent, so that there is no point of contact between him and people.[15] There is no potential for anyone, even Muhammad, to experience God. According to Islam, God did not reveal the content of the Qur'an to Muhammad, he revealed it to the angel Jibril (Gabriel in Christianese), who then revealed it to Muhammad.[16] In Christianity we share attributes with God because we are created in his image, and expressing these attributes puts us in direct contact with him. But in Islam there can be no image of God and no direct contact with God.

EXCURSUS: DO CHRISTIANS AND MUSLIMS WORSHIP THE SAME GOD?

This next part isn't really about science and religion, but it's a question that's always raised in this context. Do Christians and Muslims worship different things or the same thing? Are the adherents of one or both of these religions worshiping something other than God and thus engaged in idolatry? Or are they worshiping the same God but one or both of them is mistaken in doctrine? Religious people often think it's obvious that other religions aren't worshiping the same God as them, but nonreligious people often think it's obvious that they are. Christianity and Islam are both Abrahamic religions, after all. Fortunately, I am here to finally and definitively answer this question: I don't know.

To be more specific, I don't think *anyone* can know. To answer this question, we would have to know the answers to several other questions that have to do with philosophy of language and theories of reference.[17] I read an illustration of the difficulty here somewhere: Imagine you and I go into a crowded party in a huge venue. Then I tell you, "The guy drinking champagne and wearing the tuxedo is my father-in-law, let me introduce you." So we weave our way through the crowd to the other side of the ballroom or gymnasium or whatever but when we get there, we see that he isn't drinking champagne, he's drinking 7-UP. And he isn't wearing a tuxedo, he's wearing one of those ridiculous tuxedo T-shirts. So I gave you two referents, two ways to identify this person as my father-in-law, and both of them were wrong. *Yet it's still my father-in-law.* Despite getting

---

15. Fakhry, *History of Islamic Philosophy*, 234–35.
16. Sahih Bukhari 1.1.3–5.
17. Michaelson, "Reference."

every single referent wrong, the person I was identifying with them is identical to my father-in-law. When we apply this analogy to Christianity and Islam, we would say that just because one or both of these religions have incorrect ideas by which they refer to God (referents), it doesn't mean that they're *not* referring to God, in which case they could both be referring to the same thing, despite their differences.

But of course, we can flip the whole thing around. Same situation: Crowded party, champagne, tuxedo, father-in-law, "let me introduce you." But this time, when we get there, we see that he *is* drinking champagne, and he *is* wearing a tuxedo. But it's not my father-in-law. I used two referents to identify the person as my father-in-law; both referents were correct; but the person I was identifying with them is not identical to my father-in-law. Applying this analogy to the question at hand, we would say that just because both Christianity and Islam share correct ideas by which they refer to God, it doesn't mean they are referring to the same thing.

So which of these two scenarios is closer to the issue of whether Christians and Muslims worship the same God? Do they worship the same being despite their differences, or do they *not* worship the same being despite their similarities? And the answer, again, is: I don't know. Unless God tells us, I don't think it's *possible* for us to know. In the Bible, at least, we see both scenarios. Job's friends had false beliefs and Job is told to pray on their behalf, not because those beliefs referred to something other than God, but because they were false beliefs *about* God (Job 42:7–8). On the other hand, during the Exodus, the Israelites made a golden calf and claimed that it was the god that brought them out of Egypt (Exod 32; Acts 7:39–43). Here, just because they ascribed something to this idol that was true of God, it doesn't mean that they were worshiping God and merely got some details wrong, as with Job's friends. It's not obvious whether Christians and Muslims could say the other group is worshiping the same thing (and vice versa), despite the (allegedly) incorrect beliefs, or is not worshiping the same thing despite the correct ones.

IV. DO WE NEED A GROUND OF RATIONALITY?

Rational minds

Returning to the actual discussion, perhaps the problem with Islam and Ockhamist Christianity is that they have an element that *prevents* the universe and human beings from being rational; or at least, makes it improbable, viz., a God that *invents* rationality rather than being *intrinsically* rational. But what if we just posit a view that doesn't bring God into the equation at all? Why can't there be worldviews that don't give us a positive reason to trust our beliefs that the world is understandable and we can rationally comprehend it, but don't give us a reason to deny it either? I mean, evolution says that the fittest survive, and organisms that had false beliefs would be less fit than those with true beliefs. That should be enough, no?

No. I've already written a book on that (two, actually, but who's counting?).[18] Briefly, in order for beliefs (or, more properly, belief-forming attributes) to be selected in the struggle for survival, they would have to cause behavior by virtue of their contents—in fact, by virtue of the *truth* and *relevance* of their contents. But if the physical universe is all that exists, not only do we not have a reason to accept this, we *have* a reason to *not* accept it.

First, if nature is all that exists, how could there even be beliefs at all? I mean, beliefs have content: my belief that Antarctica is the world's largest desert has, as its content, the proposition "Antarctica is the world's largest desert." But physical objects and processes do not have content as such.[19] We can certainly ascribe content to them—we can arrange detritus on a beach to spell out SOS or something—but in and of themselves, that detritus does not have that or any other content, it's just stuff on a beach. It can *indicate* something, at least to a mind—smoke indicates fire—but the smoke doesn't have any propositional content, like "Hey, look at me! I'm on fire!"[20]

Second, if we ignore this and say beliefs can exist in a naturalistic universe, there still isn't a way for the contents of that belief to cause behavior; and that's significant because behavior is what natural selection is looking for. It selects survival-enhancing behavior or the predilection to engage in such. In philosophy, the mind-body problem considers the question of how mental properties can produce physical effects.[21] For our focus, the question is how beliefs can cause behavior by virtue of their content; and most philosophers of mind, or at least naturalist philosophers of

---

18. Well, me, I guess. *I'm* counting. I just counted to two. That counts as counting.
19. Jim Slagle, "Yes, Eliminative Materialism Is Self-Defeating."
20. Jim Slagle, "Indicators and Depictors."
21. This is another one of those endlessly referenceable issues. On one side, I recommend everything on the subject by the great Jaegwon Kim, and on the other I recommend everything on the subject by Alvin Plantinga and J. P. Moreland.

mind, say that it can't. At best, what causes the behavior is the underlying or subvening neural structures. Those are physical objects and engage in physical processes, and so they can produce a physical effect. But the belief contents that supervene on them, like "Heavy metal is a descendant of classical music,"[22] do not and cannot. And this means that the beliefs in question have no control over them. As long as the neural structure and processes are producing the effect, then it doesn't matter what the content of the belief is that supervenes on them. You see a predator and your neuronal processes cause you to hide, but the beliefs associated with those processes could be "Muenster cheese is an effective antiperspirant." You'll engage in the same survival-enhancing behavior because the neurons causing the actions are the same regardless.[23]

Third, if we ignore *this* and say beliefs do cause behavior and can be selected in the struggle for survival, this doesn't give us a reason to think the beliefs are *true*. Hiding from a predator will increase your evolutionary fitness, but any beliefs that cause that behavior would be just as liable for selection. Maybe you see the predator and think, "Look, a big bowl of ice cream! But I'm on a diet so I should keep away from it to avoid temptation." Or "I want to die and the best way to die is to hide from predators." Or "The appearance of that predator means I just won the sweepstakes, so I'd better hurry home and wait for the phone call." The point being that the number of true beliefs that produce an adaptive behavior is dwarfed by the number of *false* beliefs that produce that same behavior. Given this diversity, we have a reason to think that our belief-forming attributes, selected by the evolutionary process, would not be reliable.[24]

Of course, our belief-forming attributes *are*, by and large, reliable. The claim being addressed here does not challenge that, it's arguing that if the physical universe is all that exists, we have a reason to think they are *not* reliable, so, therefore, the physical universe is not all that exists.[25] If Judaism or Christianity or both are true, then, supposedly, we have a reason to trust our belief-forming attributes in the main, and obviously this includes scientific beliefs and reasoning. If God does not exist, or if he exists and is more like the God of Islam or Ockham, then we have a

---

22. This is true. Vivaldi is metal. Paganini is metal. Chopin and Rachmaninov are metal.

23. Plantinga, *Warrant and Proper Function*, 223–24.

24. Plantinga, *Warrant and Proper Function*, 225–26.

25. Jim Slagle, *Epistemological Skyhook*, chs. 4–5, 10; *Evolutionary Argument Against Naturalism*, chs. 5–8.

reason to *distrust* our belief-forming attributes, including scientific beliefs and reasoning.

Rational universe

The other side of this issue is why we should expect the universe itself to be rational, without a rational Creator. If there's nothing that transcends the universe and keeps it ordered, and can be trusted to do so into the future, then our beliefs that the universe is and will continue to be ordered are unfounded. If the response is that it's just the universe's nature to be ordered and structured, we can ask why *that* aspect of the universe can be trusted to continue into the future. After all, the universe is contingent: it doesn't have to have the particular properties it has; it doesn't even have to exist at all. When we ask the same thing of God, the answer is that if God exists, he is not contingent but metaphysically necessary, so he can't not exist and can't not have the properties he has. We can trust that a rational God would create an ordered universe, and we can trust that a metaphysically necessary God will keep it ordered into the future because something can't be metaphysically necessary and also be subject to change.

On the metaphysical side, this is essentially Thales's question, the first philosophical question:[26] Why do things hold together? Why is there unity instead of plurality? Why is there order instead of chaos? On the epistemological side, this is the problem of induction that Hume applied to everything but miracles.[27] How do we know the future will resemble the past? How do I know that if I stick the lit end of cigarette into my ear it will hurt? Because it hurt the first hundred times I did it? That, says Hume, doesn't follow. Induction is fallible: sometimes it works and sometimes it doesn't; sometimes it's obvious whether it will work and sometimes it isn't.

So we're left with three options: 1) Reject induction and, along with it, science and any reasoning that allows us to move, breathe, and think—including the thinking that led us to reject induction in the first place. 2) Grant the validity of induction without having any metaphysical basis for doing so. This grants the overwhelming intuition we have in favor of induction but makes it into an act of blind faith. 3) Formulate a metaphysic that gives us a reason for thinking the future truly will resemble the past

26. See above, 178–79.
27. See above, 198–99.

and that induction is valid. The existence of a God in whom intellect is prior to will, which is the ground of rationality, morality and reality, and which is metaphysically necessary provides just such a metaphysic. And the good news is we don't even have to formulate it: it was already formulated long ago before anyone thought about how we can justify induction. My point here is not to defend theism but to say that the third option—in which certain religious beliefs safeguard induction, including scientific induction—is a viable alternative. It's not irrational.

Concerns about how we can trust our conceptions of the physical world without God ultimately led Stephen Hawking and Leonard Mlodinow to affirm that the physical world does not have objective existence apart from our beliefs about it.[28] They couldn't make sense of an ordered, objective universe existing without God so they chose to reject that an ordered, objective universe exists. And if that's where your worldview leads you, then you don't get to claim that religion and science are at odds.

---

28. Hawking and Mlodinow, *Grand Design*, 172.

Conclusions

A NEIGHBOR OF OURS is going through nursing school, and one of her textbooks mentions the Latin slogan *Ubi tre physici, due athei*, which roughly translates to "Where there are three doctors, there are two atheists"—meaning that those who are steeped in medical science tend to disbelieve in God.[1] The textbook ascribes this phrase to the ancient Christians (you know, the same people who invented hospitals),[2] but it looks like it first appeared in the seventeenth-century text *Religio Medici* by Thomas Browne. Browne treats it as a common slogan of the time, and one of its later editions qualifies this by saying it is only so among the unlearned, which would be odd as Latin was used primarily by the educated elite. As to whether it was a slogan at all, in any language—well, sure, why not? But, assuming this was the case, what can we conclude from it? That it was actually *true* that seventeenth-century physicians tended to be less religious than the general populace? Or at least that they often gave the impression that they were not religious even if they were? And if we grant this, can we conclude that they were representative of scientists in general? Or that those physicians who didn't believe in God were *right*? Why think this? Why think any of this?

I don't see how we can conclude much of anything from it: it's a slogan, a catchphrase. It seems to me to be on the same level as cow tipping or porcupines shooting their quills. I emailed James Hannam about this phrase, and he responded, "It seems to have become a medieval proverb sometime in the nineteenth century. What precisely this makes it evidence of, I'm not sure. If I had to guess, I'd say it originated at the

---

1. Porter, *Blood and Guts*, 33.
2. Ferngren, *Medicine and Religion*, 113–15.

universities when the theologians and physicians found themselves in the same pub, so insulted each other in Latin!"

What interested me about this "Where there are three doctors, there are two atheists" catchphrase was my reaction upon hearing it. I immediately formed the belief that I had gone astray in thinking that the history of medicine was not characterized by hostility from Christians. Then I generalized this to all science: somehow, this one little factoid indicated that the history of science and religion really is characterized by conflict after all, and I had tricked myself into thinking otherwise. (This all took place over the course of about one second.) Then I got control of myself. While I researched it, I still had fleeting thoughts that my biases had prevented me from seeing an obvious history of how science and religion have been at each other's throats through the years. That's how strong a hold the Conflict Myth has on me, and I've taught the subject and know perfectly well that it's not true.

I. SCIENTISM

Many people say science is neither intrinsically good nor intrinsically bad, it is neutral. It's what we do with it that's good or bad. I strongly disagree with this. Science is intrinsically good. I believe this because I believe *knowledge* is intrinsically good and science is the greatest pathway for acquiring knowledge ever discovered by human beings. This doesn't dispute that we can use science for bad or evil purposes, only that this doesn't impugn science's intrinsic goodness. Just about any good thing can be used for bad purposes.

Unfortunately, some people take this to indicate that the *only* knowledge or trustworthy information we can have comes from science. By this they can mean either that anything not discoverable by science is forever unknowable, or that anything not discoverable by science does not exist—making science omniscient. They point out that scientific thought and any scientific worldview radically differ from, e.g., religious thought and worldviews. The core of the former is that any claim is falsifiable (that is, there are objective tests to determine whether it's true or false) and no claim is unchallengeable (nothing must be accepted as certain). Since religious thought doesn't have the same structure, it can never aspire to knowledge.

But this is a problem only if we *start* by taking scientific thought as exhaustive so that there is no other source of knowledge available. This, as mentioned before, is commonly called *scientism*, and since the term "scientist" is already taken, a person who advocates scientism would have to be called something like a scientisticist or a scientistician. To describe things from the perspective of scientism would be to describe them scientistically. And to translate something into this mode of thought would be an act of scientisticalization. Oh, this is fun.

There are multiple problems with scientism, however. Perhaps the most obvious is that the claim "only scientific claims are trustworthy" is not a scientific claim. It's a philosophical claim *about* science. As such, scientism is self-defeating, it is incompatible with itself. Another problem is that scientism doesn't allow for an awful lot of the knowledge we have. We know, for example, that other minds exist, that other people are sparks of self-consciousness, but we can't prove this scientifically by just observing other people's behavior or even their physical brains. Yet your friendly, neighborhood scientisticist would not be willing to give up belief in other minds—and if they *were* so willing, it would be an act of extreme irrationality on their part, not a rational one. Solipsism, the belief that one's own mind is all that ultimately exists, can be accepted only by a profoundly irrational mind.[3] A third problem with scientism is that science must make presuppositions that are metaphysical in nature, not scientific.[4] So if scientism were true, it would require us to reject the only conditions under which science could be valid, thus undermining science and scientism in one fell swoop. A fourth problem is that it is anthropocentric: human beings determine what counts as real and what isn't according to whether it's accessible to our scientific methodologies. One could claim that we're appealing to an ideal science, not just the science currently available to humanity, but that wouldn't rule *anything* out, including science discovering that other ways of discovering truth are valid and trustworthy.[5] It would also be an appeal to something that does not and cannot exist, making it, once again, inaccessible to science and scientism.

Anyhoo, these scientisticians contrast the beliefs we acquire from science and reason with the beliefs we acquire from faith. Faith, they think, means believing something in the absence of any reason

---

3. Squires, "Solipsism"; also, *please* read Fredric Brown's short story "Solipsist."

4. Trigg, *Beyond Matter*.

5. Which is not to say that we *need* science to authenticate them before accepting them.

or evidence for it—or even when there's reason or evidence *against* it. But as I mentioned in chapter 12 when discussing Tertullian, this is the definition of *blind* faith (or fideism), which traditionally is condemned in Christianity.[6] Faith means trusting God. In the same way, when you trust your parents or your spouse or whomever, you have faith in them. And not only is there nothing in this concept that's incompatible with reason or evidence, in point of fact it requires them: you wouldn't trust someone unless you had a reason to do so. It might be a *bad* reason, or the person might have given you a false impression of who they are, but without a reason you wouldn't have faith in them. Otherwise faith would just *be* blind faith, and the fact that they had to qualify the term with an adjective demonstrates that the two concepts are not identical.

Certainly, faith does not operate the same way science does in acquiring beliefs, since your reason for trusting someone may be personal or not publicly available; but no one is suggesting otherwise. "Faith . . . is the art of holding on to things your reason has once accepted, in spite of your changing moods. For moods will change, whatever view your reason takes. . . . The battle is between faith and reason on one side and emotion and imagination on the other."[7] I realize this definition doesn't even address when faith and reason seem to say contrary things,[8] which is when all the vigorous harrumphing begins, but then this book has been an attempt to show some of the most common examples of that don't hold water.

II. IDIOTS AND INTELLECTUALS

One reason some people—intellectuals in particular—can't even consider Christianity as a live option is because they think faith allows people who are not intellectuals, and even people who are *stupid*, to be justified in their confidence that Christianity is true. Science is open to everybody but it requires intellectual fortitude, and some people just don't have it in them. Whatever method allows dumb people to be rightly confident in their beliefs must be invalid. I've heard people express this by asking why

---

6. Amesbury, "Fideism."

7. C. S. Lewis, *Mere Christianity*, 123, 122; cf. "Religion: Reality or Substitute?" in C. S. Lewis, *Christian Reflections*, 37–44 (esp. 42–43).

8. A good place to start exploring this would be "On Obstinacy in Belief," in C. S. Lewis, *World's Last Night*, 13–30.

we should trust the religious experiences of a tribe of desert nomads four thousand years ago.

Note the elitism in all this. "If God exists he wouldn't reveal himself to a bunch of ignorant savages, he'd only reveal himself to people . . . like me! He'd make knowledge of himself only accessible via rationality and so only available to people who are really smart and edumacated . . . like me!" But why *wouldn't* God reveal himself to a tribe of desert nomads four thousand years ago? For that matter, why wouldn't he be just as accessible to people who are genuinely stupid as to people who are smart? Christianity claims that God is love itself. So would love refuse to express itself towards unintelligent people? To ask that question is to answer it.

So some people—not all, certainly, not even most, but *some*, especially those who try to use science as a foil to religion—can't consider Christianity as a genuine possibility because they think so many Christians are just dumb. And obviously plenty of Christians *are* dumb. Plenty are painfully naïve, plenty are militantly ignorant and closed minded, plenty use their beliefs to sow division and even hatred. And of course—*of course*—all these categories are true of atheists as well, not to mention the adherents of every other worldview. These are *human* traits, not religious or Christian traits. We all suck.

On the other hand, plenty of Christians are brilliant and more educated than most of us can ever hope to be. Thomas Nagel, an atheist philosopher, wrote how he was "made uneasy by the fact that some of the most intelligent and well-informed people [he] know[s] are religious believers."[9] Most of us, Christian and non-Christian alike, are somewhere in between scary smart and disturbingly dumb. If religion were available only to the smart people, then it *wouldn't* be available to everyone else, and the smart people could, and probably would, use that to oppress the others (being smart and being moral are not the same thing). You may want your worldview to be accessible only via a complex 130-premise syllogism that prevents dumb people from participating, but if the God of the Bible exists, he'd make himself available to everyone regardless of the score they got on an online IQ test.

In fact, if you'll recall, he often uses the most insignificant or debased elements to be the instruments of his revelation and grace. "Not many of you were wise by human standards; not many were influential; not many were of noble birth. But God chose the foolish things of the world to

---

9. Nagel, *Last Word*, 130. I feel the same unease going the other way.

shame the wise" (1 Cor 1:26–27). And that's pretty much the scenario I'm describing: some smart people are unable to consider Christianity a genuine option because foolish people believe it; and those smart people just cannot bring themselves to even entertain the possibility that they might be wrong and the fools might be right—and not by accident. It would shame them.

III. RETURN OF THE ~~KING~~ ~~JEDI~~ ~~NATIVE~~ ~~LIVING DEAD~~ PRODIGAL SON

Certain scientific developments present a new picture of the world, and since they include all the information of the old picture but correct and move beyond it, it becomes almost inconceivable that we would go back to that earlier view.[10] Once we've moved beyond geocentrism to heliocentrism, we can't seriously imagine that the geocentric view was, after all is said and done, correct, and that we should return to it. In the same way, we can't imagine abandoning germ theory in favor of humoral theory, or Newton for Aristotle, or Einstein for Newton.

I suspect this is how many people see science and religion. Science has presented a world very different than the one we thought existed, and some tie that old conception of the world to religion; in which case, going back to the religious view is as absurd as readopting geocentrism. It wouldn't merely be irrational, it would be inappropriate, unseemly, *ugly*. I've argued that many of the refuted ideas are not intrinsic to Christianity so it is an error to discard the latter just because we've discarded the former. But this idea raises a further question: Are all returns to one's beginnings unaesthetic and ugly? Some are, certainly, but some aren't. Some changes are overwhelmingly beautiful, rational, and apropos specifically *because* they are going back to what was dismissed before.

This confusion is made when people mistake their personal experience as indicative of a widespread movement. There are plenty of people who were raised in religious homes, but as they got older they found the kiddie theology they had imbibed was not very intellectually rigorous. If they explored it, all the responses to their objections seemed contrived because they were assuming the dumbed-down version they knew was the ground level, and the deeper doctrines looked like pathetic attempts to shore up the weaknesses. The idea that the more rigorous theologies

---

10. Cf. Nagel, *View from Nowhere*, 74–77.

were the ground level and the versions they heard were simplistic because, being children, they had been simplified for them either doesn't occur to them or doesn't seem particularly likely. Since this was *their* experience, it seems reasonable that's how religion has worked in general: people had stupid ideas, childlike ideas, and then when smarter people (like them) started asking inconvenient questions, the religious folk frantically scrambled around to try to find some loophole to justify the stupid ideas in an ad hoc manner. Since it would be just wrong and ugly for them—the smart people—to return to (their childhood understanding of) Christianity, that must be true of everyone else and society at large too.

But again, a return to an earlier way of thinking is often very reasonable, very aesthetically appealing, very parsimonious, very beautiful. In *The Pilgrim's Regress*, C. S. Lewis charts his intellectual development in an extended analogy in the manner of John Bunyan. Lewis was a Christian as a child but abandoned it as a young adult. Similarly, the main character of *The Pilgrim's Regress*, named John, is brought up in the land of Puritania, at the base of a tall mountain. Death means going over the mountain to the other side. John is brought to a Steward (a priest) and told about the Landlord (God). Here, Lewis brilliantly represents a child's impression of Christianity by having everyone put on a mask whenever they talk about the Landlord, showing it to be artificial and disingenuous. Then the Steward gives John a list of rules to obey—"but half the rules seemed to forbid things he had never heard of, and the other half forbade things he was doing every day and could not imagine not doing." The Steward tells him that if he breaks any of the rules, the Landlord will put him in a black hole (hell). When John asks if there is any way to avoid the black hole if he'd already broken a rule, the Steward "sat down and talked for a long time, but John could not understand a single syllable. However, it all ended with pointing out that the Landlord was quite extraordinarily kind and good to his tenants, and would certainly torture most of them to death the moment he had the slightest pretext."[11] I love this.

John has a vision of an island in the west, and so leaves Puritania to pursue it. The island represents longing or *Sehnsucht*, what Lewis later calls "joy" in his autobiography.[12] The first person he encounters on his journey is Mr. Enlightenment, who greatly comforts John by telling him that there is no Landlord. When John asks him how he knows this for sure,

---

11. C. S. Lewis, *Pilgrim's Regress*, 5.
12. C. S. Lewis, *Surprised by Joy*.

Mr. Enlightenment shouts, "Christopher Columbus, Galileo, the earth is round, invention of printing, gunpowder!!"[13] Eventually, John acquires a traveling companion named Vertue, and they get to a seemingly impassable canyon, so the journey then becomes an attempt to find some way of crossing it. They travel north, where they meet nihilism, and south, where they meet philosophy. Mother Kirk (Christianity) tells them that she can carry them across, but John doesn't want anything to do with her. Finally, he despairs, dives into a pool (accepts Christianity), and swims through underwater caverns to a beach on the other side of the canyon. There, John finally sees his island, the fulfillment of everything he'd been longing for all his life, a short distance across the water. He then realizes that the landscape of the island is the same shape as the mountain in Puritania: he'd gone all the way around the world and was seeing it from the other side, arriving almost at the place he departed from. But there are no boats: the only way to get to the island is to return the way he came.

This is a deeply beautiful idea, that his attempt to get as far away from his origins as possible—and his pursuit of truth and beauty, which was essentially the same thing—led him back to the place where he began. To be sure, the way back looks very different than it had on the way out. It's a return, but it's a radically transformed return. There *is* a simplistic understanding of it all, represented by his beginnings in Puritania. But that simplistic understanding is continuous with his mature, revitalized understanding. They both end up being the same place, just seen from different angles.

Now of course, this is just one person's experience too; we can't generalize this any more than we can generalize the experience of the person who mistakes his childhood understanding of religion for the full concept. The idea isn't to generalize anything, it's that we can't paint religion so simplistically that we can simply reject it as outmoded in the face of contemporary science. Religion is much more robust than that.

So I guess my point isn't a point so much as a double-edged sword. If you find religion appealing on some level but have been unable to take it seriously because of science, then this is your lucky day: it's wide open to you. You can look at religions on their own terms. The barrier separating the world of science from religion is essentially a smokescreen. However, if you *don't* want to be religious, and you've been using science to justify this view, you're going to have to find a better excuse.

---

13. C. S. Lewis, *Pilgrim's Regress*, 21.

Appendix

The Christ Myth Myth

THIS APPENDIX IS A sort of addendum to chapter 13 about miracles. It's only tangentially about science insofar as it addresses one particular miracle claim rather than whether miracles are compatible with science. The reason I'm including it is that people will often refuse to consider the philosophical issues involved in miracles because they think there are no good examples of any (I know this from personal experience). My goal here is not to provide an example, per se, but to rebut a common nonscientific objection to the primary example of a miracle in the philosophical literature, viz., the alleged resurrection of Jesus, so that the issues in chapter 13 can be tackled. Moreover, it fits into the larger project regarding urban legends about religion and Christianity that intellectuals commonly believe.

Virtually every culture in history has had a creation myth that is often moderately similar to the biblical story of Adam and Eve in the garden of Eden. Additionally, many cultures have had a flood myth like the story of Noah, where a flood wipes out all or most of humanity except for one man and his family.[1] This point is not really contested or controversial, although obviously it's massively controversial what it *means*: whether we should conclude that the creation and flood stories in the Bible are myths or folk stories or parables or some other form of nonhistorical literature. But I was weaned on an idea called the *Christ Myth*. This suggests the biblical account of Jesus is in the same category as the creation and the flood

1. Dundes, *Flood Myth*.

stories, in that there are parallels to Jesus' life, death, and resurrection throughout the religions and mythologies of the world, implying that the Jesus story is neither unique nor authentic.

The claim here is not about broad ideas that religions tend to share, like gods, the supernatural, and the afterlife, or the problems of (and solutions to) human nature, since it's no surprise that religions have these elements. It's that we find more specific parallels like baptisms, Last Suppers, virgin births, and resurrections all over the place. So why take the Christian versions of these stories as true and historical? If we can (*if* we can) reject the creation and flood accounts as nonhistorical parables, why can't we read the Gospels the same way?

There are several responses to make to this charge. First, though, I should point out that when people correct important beliefs they sometimes overcorrect, and their biases flip from one end of the spectrum to the other instead of to a more balanced position. Since the Christ Myth is where I started from, I fear this may be true of me. I'm not suggesting you dismiss what follows because of this, just bear it in mind.

I. THE ALLEGED PARALLELS

Essentially, there are two claims made. One is that Christianity borrowed its specific ideas from the mystery religions or Gnosticism that existed in the Roman Empire. The second is broader: that similar ideas exist in myths and religions all over the world, implying that we naturally put things into those categories associated today with Christianity. Our minds are hardwired to believe in resurrections, virgin births, et al.

Some noteworthy people have not only accepted these parallels but thought they made Christianity *more* plausible, not less. C. S. Lewis, for example, had dismissed Christianity as a young man, but eventually accepted it and claimed the universal dying-and-rising-god story was one of his main reasons for doing so. In Jesus "myth became fact."[2] Basically, he took these stories as God preparing us to accept the claims of Jesus and his offer of salvation by providing us with a ready-made context for them.[3]

There's a slight problem with this though. The idea that there are parallels, especially in the Near East, to Jesus' birth, life, death, and

---

2. He makes this point in several places, but the most obvious is "Myth Became Fact," in C. S. Lewis, *God in the Dock*, 63–67.

3. See also Lapide, *Resurrection of Jesus*, 120–22.

resurrection had its heyday with the *religionsgeschichtliche Schule* in Germany between about 1890 and 1920. The reason the movement died out was pretty basic: *the parallels were bogus*.[4] The virgin births weren't virgin births, the resurrections weren't resurrections. So the Christ Myth is itself a myth: it's not real.[5] It was never anything but a marginal hypothesis, held by only a minority of scholars, which had no evidence in its favor, and which was discarded a century ago.

To make the alleged mythological parallels to Jesus seem parallel in the first place, they had to describe them with Christian terms and categories—then they'd turn around and marvel at the similarities.[6] If you read the original source material, the case quickly collapses. For example, I've read a few dozen myths or accounts of myths that contain the "virgin birth" motif. Obviously this isn't exhaustive, but so far I can group these birth stories into four categories:

1. Virgin births in which the woman becomes pregnant by having sexual intercourse. Now I'm not a gynecologist, but I'm pretty sure the sexual intercourse part precludes it from being a virgin birth. This single category makes up the vast majority of the stories and it obviously doesn't meet the minimal criteria. It's just *weird* to call a birth that's the product of sexual intercourse a "virgin birth."

2. Virgin births in which semen is deposited near enough to a woman that she becomes pregnant. So, for example, a god leaves his "seed" in a pool, a virgin bathes in the pool, and the seed heads upstream. It makes sense to call these virgin births because the woman *is* a virgin: she has not had sexual intercourse. But it still involves all the standard props—it's just that the props have to swim a little farther than usual. It has nothing in common with Jesus' conception and birth, however, whence we get the term "virgin birth," so calling it such blurs the radical differences between them.

3. Virgin births in which the person isn't born at all but just sort of *emerges*, like Mithras coming out of a stone, or Athena springing from Zeus's forehead. To call these "virgin births" is worse than the first category since they don't even involve *births*, much less virgins. I mean, granted, Mithras's stone never engaged in any sexual activity, but if we

---

4. Rudolph, "Religionsgeschichtliche Schule."

5. Note that here and in the title of this appendix I'm using two different definitions of myth because I'm so clever.

6. Bevan, "Mystery Religions and Christianity," 43; Rahner, *Greek Myths and Christian Mystery*, 34; Nash, *Gospel and the Greeks*, 116.

start applying this category to inanimate objects, we've gone *way* too deep down the rabbit hole.

4. Virgin births in which it is not related how the woman became pregnant. I remember one account, I think it was in a Joseph Campbell book but I can't find it now, about a woman who left her tribe to go commune with the god on the mountain. She came back a year later with a baby. My recollection is that the author referred to this as a virgin birth, except the story itself wouldn't suggest it. Maybe I'm misremembering it though and *I'm* the one who applied that category to the story.

In contrast with these, the virgin birth of Jesus did not involve a man, a penis, semen, or any of the other customary accoutrements.

> Any comparison of Matthew 1–2 and Luke 1–2 to pagan divine birth stories leads to the conclusion that the Gospel stories cannot be explained simply on the basis of such comparisons. . . . For what we find in Matthew and Luke is not the story of . . . a divine being descending to earth and, in the guise of a man, mating with a human woman, but rather the story of miraculous conception without the aid of any man, divine or otherwise. As such, this story is without precedent either in Jewish or pagan literature.[7]

God supernaturally created the genetic material in Mary's uterus, combining it with one of her ova. The terms "virgin birth" and "virginal conception" are meant to refer to this entire idea, not just the part about Mary never having had sex before (which is why category 2, above, is misleading). This does not sound similar to the above stories unless you broaden the categories so much that they can apply to nearly anything. Thus, contemporary New Testament scholars do not think there are any similarities between Jesus' conception and birth and those of the various myths of the world, beyond the facts that they involve births and the supernatural.[8]

Appeals to parallel accounts of Jesus are made for the other categories too: anytime someone gets wet it's labeled a "baptism"; anything involving food and drink is a "Last Supper" or "Eucharist." The same holds true of resurrecting gods. A few decades ago the story that was held up as the most similar to Jesus (in the books I was reading at the time, at least) was Osiris. The story is that he and his sister Isis were lovers because, you

---

7. Witherington, "Birth of Jesus," 70.

8. Witherington, "Birth of Jesus." For some specific examples, see Machen, *Virgin Birth of Christ*, 323–79.

know, ancient Egypt. But their brother Typhon, aka Set, killed Osiris and sunk his body in the Nile River (sorry: "baptized"). Isis managed to find Osiris's body—most of it anyway since a fish had eaten his man bits—but then Typhon chopped up the body and scattered the pieces throughout Egypt. Isis found all the pieces, put them together along with a golden dong she'd had constructed and had sex with it.[9] And at some point, Osiris's spirit awoke in the underworld.

Aaaaaaaand scene. That's it. That's the resurrection of Osiris.[10] When I was reading those books back in the day, I thought the resurrection had something to do with Osiris's life force returning to his body long enough to impregnate Isis via his shiny, new metal wiener, but later I read suggestions that it was Osiris's "survival" in the underworld that was the resurrection (which would make any concept of an afterlife a "resurrection"). Neither case is particularly impressive as an analogue to Jesus' resurrection.

The claims I've read more recently hold up Horus, the son of Osiris and Isis, as being the closest parallel to Jesus. At some point Horus fell ill and Isis realized he'd been stung by an asp or scorpion or something. So she asked two other gods to give her magical powers to remove the poison. Basically, she did what any goddess of healing worth her salt would do: she healed him.[11] (Prayers to Isis were for protection from illness and harm.) The idea of resurrection comes in when some suggest that Horus died and Isis brought him back from the dead. But Isis was not given any such power; she had only the power to heal. Not to mention that there's no indication that Horus died in the first place. Removing poison from a dead body doesn't do anything.

To reiterate, Osiris and Horus are held up as two of the closest parallels to Jesus' resurrection and they are obviously nothing like it. Other claims are no better. "Which mystery gods actually experienced a resurrection from the dead? Certainly no early texts refer to any resurrection of Attis. Attempts to link the worship of Adonis to a resurrection are equally weak. Nor is the case for a resurrection of Osiris any stronger."[12]

9. The phallus from the palace holds the brew that is true.
10. Plutarch, "Isis and Osiris," 355D–58E (5:30–49).
11. Mary Jo Sharp, "Does the Story of Jesus Mimic Pagan Mystery Stories?" in Copan and Craig, *Come Let Us Reason*, 156–57.
12. Nash, *Gospel and the Greeks*, 161; cf. Machen, *Origin of Paul's Religion*, 234–35. For debunkings of some of the most common claims of resurrection, see Jonathan Z. Smith, "Dying and Rising Gods," and Mark S. Smith, "The Death of 'Dying and Rising Gods' in the Biblical World."

There *are* plenty of mythological stories of gods or heroes performing great actions and being transformed as a result—the transformation sometimes being called a rebirth or resurrection—which is called the Hero's Journey.[13] I'm not challenging this at all, I would only point out it is far too broad a concept to be used to dismiss a claim's historicity since plenty of historical events fit this pattern too. Besides which, what exactly would we *expect* mythological stories to be about? Mundane people doing mundane things while living mundane, unchanging lives? *Of course* myths are going to be about superhuman figures doing mighty deeds and being changed as a result. And any account of their lives will go over their beginnings (births) and endings (deaths). But again, this applies to plenty of real historical events and personages. The idea of a hero being born in squalor, rising to a position of eminence, and then freeing those who are captive and dying in doing so describes Abraham Lincoln just as much as Jesus.[14]

Moreover, all these dying god stories are tragedies. In contrast, the death of Jesus is a triumph (that's why we call the day honoring it Good Friday).

> Christianity stands entirely apart from the pagan mysteries in that its report of Jesus' death is a message of triumph. Even as Jesus was experiencing the pain and humiliation of the cross, He was the victor. The New Testament's mood of exultation contrasts sharply with that of the mystery religions, whose followers wept and mourned for the terrible fate that overtook their gods.[15]

In addition, only Jesus died voluntarily, for his followers, and for sin. There is nothing like this in any of the alleged dying god myths.[16]

I'd always read that dying and rising gods were meant to represent the vegetation cycle, where plants die in autumn and winter and regerminate in spring, but according to Jonathan Z. Smith, this "explanation is flawed at the level of theory. Modern scholarship has largely rejected, for good reasons, an interpretation of deities as projections of natural

---

13. Campbell, *Hero with a Thousand Faces*. Note that Campbell is exceptionally prone to exaggerate similarities and ignore differences in order to make the stories fit his categories.

14. I can imagine, two thousand years from now, people claiming that the Jesus story is derived from the story of Abraham Lincoln.

15. Nash, *Gospel and the Greeks*, 161.

16. Metzger, "Methodology in the Study of the Mystery Religions and Early Christianity," 18; Nash, *Gospel and the Greeks*, 160–61.

phenomena."[17] But ignoring this, trying to compare the vegetation cycle to Jesus' death and resurrection can be done in only the vaguest of senses. Lewis—who, remember, *accepted* these stories as parallel[18]—wrote,

> I myself, who first seriously read the New Testament when I was, imaginatively and poetically, all agog for the Death and Re-birth pattern and anxious to meet a corn-king, was chilled and puzzled by the almost total absence of such ideas in the Christian documents. One moment particularly stood out. A "dying God"—the only dying God who might possibly be historical—holds bread, that is, corn, in His hand and says, "This is my body." Surely here, even if nowhere else . . . the truth must come out; the connection between this and the annual drama of the crops must be made. But it is not. It is there for me. There is no sign that it was there for the disciples or (humanly speaking) for Christ Himself. It is almost as if He didn't realise what He had said.[19]

II. METHODOLOGY

So how do you get from these contrived parallels to Jesus? I'll demonstrate using Joseph Campbell, who some people hold up as an expert in comparative mythology but who was really more like a New Age guru who used comparative mythology to justify his claims.

> Thor is called, in Scandinavia, The Defender of the World, and amulet miniatures of his hammer have for centuries been worn to afford protection. At Stockholm, the museum holds one of amber from a late paleolithic date; and from the early metal ages fifty or more tiny T-shaped hammers of silver and gold have been collected. In fact, even to the present—or, at least, to the first years of the present century—Manx fishermen have been accustomed to wear the T-shaped bone from the tongue of a sheep to protect them from the sea; and in German slaughterhouses workers have been seen with the same bone suspended from their necks.

---

17. Jonathan Z. Smith, "Dying and Rising Gods," 521.

18. Although in *Till We Have Faces*, Lewis has the main character, in addressing a myth representing precipitation watering the earth, say to herself, "It's very strange that our fathers should first think it worth telling us that rain falls out of the sky, and then, for fear such a notable secret should get out (why not hold their tongues?) wrap it up in a filthy tale so that no one could understand the telling" (271).

19. C. S. Lewis, *Miracles*, 117–18.

> An unforeseen, somewhat startling overtone is added by this observation to the T-motif that has already been discussed in connection with the Celtic Christian Tunc-page (which is of a date when the Celtic and Viking spheres of influence were in many ways interlaced); and, of course, then vice versa: the apparently merely grotesque fishing episode acquires a new range of possible significance when the T of the Celtic page is identified with Thor's hammer as well as with Christ's cross. We might, in fact, even ask whether in Manx and German folklore the T-shaped bone of the sheep—the sacrificial lamb—may not have been consciously identified with the world-redeeming cross of the man-god Christ, as well as with the world-defending hammer of the native, far more ancient, even possibly paleolithic, man-god Thor.[20]

And that, ladies and gentlemen, is how you make a conspiracy theory. 1) Find some random and unrelated factoids. 2) Divorce them from the contexts in which you encountered them. 3) Construct a vague theory about how these factoids are somehow related, ignoring all the evidence that puts them in the contexts you just divorced them from. 4) Portray your contrived theory as the norm and suggest all deviations from it are what's *actually* contrived.

Campbell references A. C. Haddon,[21] who points to the sheep tongue bones worn by some people in Whitby in northeastern England, "and *probably* [my emphasis] by other Yorkshire fisherman" as good luck charms to protect them from drowning. Basically, some people in a small village wore them and Haddon thinks it probable that the practice extends beyond that village, although no reason for this extension is given. Then Haddon says one person suggested to him that this T-shaped bone *might* (my emphasis again) be meant to represent Thor's hammer, another person told him that Manx fisherman (from the Isle of Man) wore the same thing, and a third person said some Berlin slaughter yard workers "have been seen" wearing something similar. The third person also said that representations of Thor's hammer were worn in the early Iron Age in Denmark and suggests that they devolved from a fetish to a lucky charm.

The intermediate steps between silver and gold amulets in Denmark in the early Iron Age and sheep tongue bones in England in the early twentieth century go unmentioned. Nor do we hear how these amulets

---

20. Campbell, *Occidental Mythology*, 479–80.
21. Haddon, *Magic and Fetishism*, 39–40; also Phillpotts, "German Heathenism," 481–82.

could be said to represent Thor's hammer, since there is a millennium-long gap between the early Iron Age and the first references to Thor (despite Campbell's claim that Thor is "far more ancient" than Jesus). And a T-shape is a pretty simple form, after all: it's just two intersecting perpendicular lines. What about all the fishermen from other areas who use different good luck charms? What about the charms worn by farriers, farmers, and fletchers rather than fishermen? You're bound to find charms taking many diverse forms, allowing you to pick and choose the ones that let you make your point and ignore the rest as irrelevant; not to mention the focus on charms in the first place instead of repeated semi-magical phrases, hand gestures, etc. I'd like more evidence than some people speculating about a possible connection referencing each other.

As outré as this already is, Campbell shifts the whole thing from first gear into fifth. He finds some titles or descriptions of Thor that sound vaguely like they could be said of God and Jesus in the Judeo-Christian tradition. He then adds that a T-shape is basically a cross, which is obviously a major symbol of Christianity, comparing Thor's "world-defending hammer" to Jesus' "world-redeeming cross." And to tie it all off with a bow, he suggests the use of a bone from a presumably (and hopefully) dead sheep might be based on the Christian idea of Jesus as a sacrificial lamb.

Dude. Do I really have to explain how ridiculously contrived this is? You could find connections between *anything* with this kind of reasoning. Maybe the reason they were popular among fishermen had something to do with the Christian ichthus fish symbol that we see on so many cars today. Maybe it's related to the fact that Christians are called to be "fishers of men" (Matt 4:19; Mark 1:17). For that matter, maybe using rabbits' feet as lucky charms devolved from an earlier practice involving lamb's feet (because Easter switched from sheep to bunnies, you see), which was symbolic of Jesus as sacrificial lamb. I mean, you can say anything about anything this way.

That's Campbell's methodology. Find a myth wholly unrelated to Christianity: check. Find some element of that myth that can be implied to be similar to an element of Christianity when both are divorced from their larger contexts: check. Rinse and repeat.

III. THE STORY ON THE STORIES

It should have been evident that this approach to Jesus was wrongheaded from the get-go. Obviously, the proper context for understanding a *first-century Jew* is *first-century Judaism*. To think world mythology is the right milieu shouldn't even have been on the table. But this is partially due to that particular blindness of scholars: missing the forest for the trees. Fortunately, since the Third Quest for the historical Jesus began in the late 1970s, scholars have been more strongly emphasizing Jesus' Jewishness.[22]

Now, I led with the "parallels ain't parallel" case because that's the most important point. But there are other issues that make it even worse. For example, from the first through third centuries AD Christianity was rabidly anti-syncretistic, refusing to allow religious ideas from outside the faith from having any influence on Christian beliefs and practices—a continuance and intensification of Judaism's anti-syncretism.[23] Some of the mystery religions had vaguely similar rites, like rituals involving washings or food and drink, and a few church fathers took these to be imitations of the Christian practices.[24] But Judaism had these elements too and would have been the obvious source for the Christian ideas. So the whole myth conception can't get off the ground. It doesn't even make it to the runway. Really, it explodes in the hangar.[25]

Moreover, the reason the mystery religions are called mystery religions is because they are (you'll never guess) *mysterious*. We don't know much about them prior to the third and fourth centuries AD because their adherents swore vows of secrecy. In many cases we can make only educated guesses based on sculptures and bas-reliefs. Not to mention that they were in constant flux, as should be expected since they eschewed fixed doctrines, making it impossible to take something from a mystery religion in the fourth century AD and apply it to the third, much less the first.[26]

---

22. Wright, *Jesus and the Victory of God*, 13–124.

23. Metzger, "Methodology in the Study of the Mystery Religions and Early Christianity," 7; Machen, *Origin of Paul's Religion*, 238, 255–56; Hooke, "Christianity and the Mystery Religions"; Witherington, *Jesus Quest*, 80; Wright, *Jesus and the Victory of God*, 62–65; Nash, *Gospel and the Greeks*, 157.

24. E.g., Justin Martyr, *First Apology* 66; Tertullian, *De corona militis* 15; Origen, *Against Celsus* 6.22.

25. So said the great theologian Calvin. See Watterson, *There's Treasure Everywhere*, 126.

26. Metzger, "Methodology in the Study of the Mystery Religions and Early Christianity," 6–7; Rahner, *Greek Myths and Christian Mystery*, 21, 23.

There are two other points to make here that are moderately devastating: first, while the mystery religions themselves predate Christianity, almost none of the particular elements in them that are held up as parallel do. They don't appear until *after* Christianity had been around for a while. Second, even ignoring this, none of these religions was present in first-century Israel. So there's no opportunity for them to have influenced nascent Christianity, even if the Christians had been willing to adopt them.[27]

From what I've read, I understand that Christians abandoned their anti-syncretism and adapted some pagan ideas into Christianity (although it wouldn't have been until the fourth century AD), presumably to seduce people away from them. The examples seem to have been exaggerated though. Regardless, if true, it might have been a response to some of these religions adopting Christian elements for the same reason. This is where a lot of the claims that Christianity copied from them come from.[28] Nevertheless, all the alleged similarities are deeply, deeply superficial, and involved only Christian *practices*, not the initial ideas of Christianity, like Jesus' birth, life, death, and resurrection.

As already mentioned, the apex of these myth takes was the *religionsgeschichtliche Schule* in Germany, whose "representatives attempted, as those of hardly any other theological movement of the past had done, to broadcast their view on a large scale through popular works."[29] It started around 1890 and was primarily associated with one school (the University of Göttingen). The movement lost its influence after World War I, but their claims were still debated into the 1930s and even beyond. They focused specifically on Near East myths and mystery religions because they wanted to demonstrate a historical connection between them and Jesus—that the early Christians were influenced by myths they had heard and modeled Christianity after them, which we've already seen does not work on multiple levels.

So why did these German New Testament scholars in the late nineteenth and early twentieth centuries think mythology would provide the proper background for understanding Jesus rather than Judaism? Well, the fact that they were late nineteenth- and early twentieth-century

---

27. Metzger, "Methodology in the Study of the Mystery Religions and Early Christianity," 8; Rahner, *Greek Myths and Christian Mystery*, 24–27; Nash, *Gospel and the Greeks*, 117–20.

28. Metzger, "Methodology in the Study of the Mystery Religions and Early Christianity," 4–6; Rahner, *Greek Myths and Christian Mystery*, 43–45.

29. Rudolph, "Religionsgeschichtliche Schule," 12:293.

Germans tells you the answer: it was because of the rampant anti-Semitism in Germany and most of Europe at the time.[30] They didn't want a *Jewish* Jesus, they wanted a *pagan* Jesus, so they started looking through these mythologies to find something that might fit the bill.[31]

After their movement collapsed around 1920, a similar suggestion that had been made by Richard Reitzenstein was taken up and expanded upon by Rudolf Bultmann and others that Christianity copied its ideas from a Gnostic Redeemer Myth. But this was even worse.[32] Gnosticism itself dates only from the second century AD, after all, and Christianity had already become widespread by then, so any borrowing would have been *from* Christianity, not *by* Christianity.[33] The gnostic redeemer theory relied on the Hermetic, Nag Hammadi, and Mandaean literatures, all of which *postdate* Christianity, sometimes by several centuries, and then tried to read them into the New Testament.[34] Not to mention that the myth in question does not unequivocally involve death and resurrection in the first place: a redeemer descends from heaven with spiritual knowledge (*gnosis*) that the sect must keep secret and then ascends back to heaven. While Bultmann had an enormous influence on Protestant theology in the first half of the twentieth century, this theory has been almost entirely rejected at least since the Third Quest began half a century ago.

Despite the collapse of the *religionsgeschichtliche Schule* in the early twentieth century, the idea of mythological parallels took off all over the rest of academia, notably in the social sciences, to the point where it seemed to be almost axiomatic.[35] In the late nineteenth and early twentieth centuries, anthropologist James George Frazer published his multivolume *The Golden Bough*, which pointed to similarities in various

---

30. As opposed to today where the rampant anti-Semitism is qualified with "You know I'm not anti-Semitic but . . ."

31. Please note this isn't an argument against their claims of parallel myths since that would be an ad hominem fallacy. Nor am I suggesting that those who promote the Christ Myth today have similar motivations.

32. Nash, *Gospel and the Greeks*, 191–246.

33. Yamauchi, *Pre-Christian Gnosticism*; Dunn, *Beginning from Jerusalem*, 41.

34. See Yamauchi, *Pre-Christian Gnosticism*, 69–72 (Hermetic), 101–16 (Nag Hammadi), 117–42 (Mandaean), and Nash, *Gospel and the Greeks*, 217–21 (Mandaean), 226–35 (Hermetic), 236–40 (Nag Hammadi).

35. "The Religionsgeschichtliche Schule began as a movement within theology, but it ended outside theology because its methods and approach were so radical" (Rudolph, "Religionsgeschichtliche Schule," 12:295).

mythologies of the world. He acknowledged his case was speculative and based only on circumstantial associations and vague resemblances,[36] but the confidence others had (and have) in his speculations was astounding. He often referred to the dying-and-rising-god motif. But the category for rebirth or resurrection was so broad it could apply to just about any change, which is, in a sense, "dying" to whatever was before the change and being "resurrected" or "reborn" in the resulting state.

> The category of dying and rising gods, once a major topic of scholarly investigation, must now be understood to have been largely a misnomer based on imaginative reconstructions and exceedingly late or highly ambiguous texts. . . . All the deities that have been identified as belonging to the class of dying and rising deities can be subsumed under the two larger classes of disappearing deities or dying deities. In the first case, the deities return but have not died; in the second case, the gods die but do not return. There is no unambiguous instance in the history of religions of a dying and rising deity.[37]

Similarly, the psychologist Carl Jung suggested there were psychological archetypes that form our stories, and some have argued these archetypes are expressed through the mythologies of the world. I can't say anything for or against Jungian psychoanalysis in general, but his theory of archetypes is too abstract to be susceptible to empirical testing.[38] Besides which, you don't find the categories of Jung (or Frazer) in many of the largest religions of the world. Using them to explain away religious beliefs ends up being just a matter of picking and choosing what you want to include, which is precisely how conspiracy theories operate.

But this still doesn't answer the question why the idea of mythological parallels to Jesus was so attractive to so many people that it became all but indisputable despite being false, even today. In 2007 a film came out called *The Man from Earth* with a main character who's been alive for thousands of years, and has been numerous figures in history, including Jesus. The other characters then contemplate how the elements of the

---

36. Marrett and Penniman, *Spencer's Scientific Correspondence*, 22; Downie, *Frazer and the Golden Bough*, 112; Cawte, "It's an Ancient Custom," 38. See also Fontenrose, *Ritual Theory of Myth*.

37. Jonathan Z. Smith, "Dying and Rising Gods," 521, 522. I presume Smith excludes Jesus from his pronouncement that "there is no unambiguous instance . . . of a dying and rising deity," but I don't want to assume too much.

38. Roesler, "Are Archetypes Transmitted More by Culture than Biology?"

Jesus story are present in many religions and mythologies of the world.[39] That same year, a documentary called *Zeitgeist* was released that defended the Christ Myth and tied it to 9/11 conspiracy theories and bankers trying to control the world.[40] So, again, why is a theory that was only ever held by a small minority of scholars over a century ago, which never had any evidence in its favor and has been decisively debunked and rejected by historical Jesus scholars, still have such a draw on Western culture that people unquestioningly accept it a hundred years after it went belly up?

The specific motif of death and rebirth/resurrection is probably a holdover from the attempts to treat evolution as a grand unified theory of everything that applies to all subjects, whatever their distance from (or relevance to) biology may be. The evolution of one species into another can very easily be seen as the first species dying and being resurrected in a new form or body. As discussed in chapter 8, this mission to see everything through Darwin-colored glasses was primarily a phenomenon of the late nineteenth and early twentieth centuries, concurrently with the *religionsgeschichtliche Schule*, although it still goes on today.[41] So seeing death and resurrection written into the very fabric of the universe led some people to use it to try to explain away the resurrection of Jesus.

But this is only one specific motif, and the evolutionary explanation of it doesn't apply to the other cases. What about all the erroneous claims of virgin births, baptisms, and the like? I can only follow Frazer and Jung by speculating about this. I think it was because it just *felt* right: the Christian story was merely one expression of the universal story,[42] written into our very psyches, thus allowing us to forswear Christianity and pursue the universal story instead—or reject it because it doesn't matter whether you accept it or not. A similar effect took place, I think, with the Rutherford-Bohr model of the atom, which portrayed the atom kind of like a solar system, the nucleus taking the place of a star and the orbiting electrons taking the place of planets. It didn't take much

---

39. One of the other historical figures the man claims to have been was a sailor on Columbus's first voyage across the Atlantic. The filmmakers take this opportunity to put forward the whole flat-earth canard (see ch. 2).

40. Mark W. Foreman, "Challenging the *Zeitgeist* Movie: Parallelomania on Steroids," in Copan and Craig, *Come Let Us Reason*.

41. "Evolutionary hand waving is an example of the tendency to take a theory which has been successful in one domain and apply it to anything else you can't understand—not even to apply it, but vaguely to imagine such an application" (Nagel, *View from Nowhere*, 78).

42. Cf. Vainio, *Cosmology in Theological Perspective*, 54.

prompting for some non-physicists to speculate that each atom may actually *be* a solar system, and our solar system may be an atom of a larger reality. It's a beautiful idea, it's just wrong.

It's also similar to Ernst Haeckel's embryology: as each individual organism develops in the womb or egg (ontogeny), it goes through (recapitulates) the evolutionary stages of its species (phylogeny). The development of each individual organism is a microcosm of its species' evolutionary history. So human fetuses start off looking something like fish, then amphibians, then reptiles, etc. To make the case, however, Haeckel had to severely exaggerate the similarities between the various fetal stages and some animals in that species' evolutionary development. This idea has also been rejected for well over a century but, like the Christ Myth, it was accepted for decades by nonbiologists within academia and without, and sometimes still is. It's a very poetic idea, an organism reliving the evolutionary development of its entire species. It feels like it *should* be right. But it isn't.

The irony is that many of the people who accept the alleged parallels to Jesus present themselves as well-educated skeptics. This reminds me, again, of how some scientists were so gullible that they believed a silly story of a pope excommunicating Halley's Comet because they thought it showed how gullible *other* people were.[43] In the same way, those who buy in to the claim that there are parallels to Jesus in world mythology are evincing a gullibility far exceeding what they are accusing Christians of. If they expressed a tenth of the skepticism to their own claims as they did to the Jesus story itself, they would never have accepted the Christ Myth in the first place.[44]

IV. THE PROBLEM OF LITERARY GENRES

Now, when people call the Gospels "myth," sometimes they mean only it's not true. But if they're comparing it to mythological stories, there's

---

43. See above, 8–11.

44. See, e.g., Evan Fales's treatment of this in his "Successful Defense?" He says he is an unabashed skeptic (7), but then tries to explain away the Gospels via "the analysis of myths and myth construction" (28), and Jesus' resurrection in particular by comparing it to "resurrection traditions" (27) in the ancient Near East "of which there is no dearth of examples" (29). Thus, he blindly avows the long-debunked idea that there are mythological parallels to Jesus' resurrection with a gullibility that is truly stunning—all in the name of skepticism. It's frustrating because I *like* Fales, he is an otherwise outstanding philosopher and thinker.

another problem. Mythology is, among other things, a family of *literary genres*. If you read an actual myth, not just a modern summary of one, it becomes pretty obvious pretty quickly that they have a general tone to them. By contrast, other literary genres, like historical writings, do not read like myths. I'm contrasting mythology with history because sometimes historical accounts develop into mythological accounts, and you can tell roughly how far along the mythologization process a story is via literary analysis.

Professional lunatic Tim Severin once built and sailed a leather boat from southern Ireland to the Faroe Islands, then Iceland, and then all the way across the ocean to Newfoundland. He got the idea from his wife, who was reading *The Voyage of Saint Brendan*, a ninth- or tenth-century text that describes an Irish monk from the late sixth century building a currach out of leather and sailing it around the North Atlantic. Some of the accounts are pretty fanciful, but there are other details that you don't usually find in mythology. Severin's wife told him that it sounds like an account of real events that was in the process of being mythologized.[45] There were a lot of details that didn't add to the story, for example. In mythology you're trying to make a point, and irrelevant details distract from it, so they tend to be weeded out. Historical accounts are constricted by the actual events and so tend to have more details that don't go anywhere. (This is just a general characteristic, there are plenty of counterexamples.) Some people have suggested that the descriptions in *The Voyage of Saint Brendan* sound like he might have sailed (in the sixth century, remember) all the way to North America in a leather boat—hence Tim Severin trying to see if it could be done. Severin was successful, but I make no judgment about Brendan.

A moderately well-versed reader can often tell whether a text is trying to communicate real historical events or nonhistorical stories, although there is plenty of gray area, as the case with Severin's wife shows (which is why we need historians to sift through the texts to determine which parts are historical and which aren't). But it's only in the modern era that literary analysis has given us the resources to analytically distinguish different genres of literature. And the Gospels are not written in a genre of mythology, they are written in a genre of historical writing. C. S. Lewis, who was a literary expert, wrote of the Gospel of John (the one most accused of being nonhistorical), "I have been reading poems, romances, vision-literature, legends, myths all my life. I know what they

---

45. Severin, *Brendan Voyage*, 12–13.

are like. I know that not one of them is like this."[46] In fact, we can make a stronger and more specific claim: for the last few decades scholars have recognized that the Gospels are written in the historical literary genre of *Graeco-Roman biography*.[47]

This creates a problem. Those who say we should read the Gospels as mythology do not say we should do so because they are *written* like myths but because they contain mythological motifs, like virgin births and resurrections. We've already seen that's a dead end. But ignoring that, what would it mean to say people in the first century wrote myths or any nonhistorical stories *as if* they were historical? Once again, Lewis provides the answer:

> Of this text there are only two possible views. Either this is reportage—though it may no doubt contain errors. . . . Or else, some unknown writer in the second century, without known predecessors or successors, suddenly anticipated the whole technique of modern, novelistic, realistic narrative. If it is untrue, it must be narrative of that kind. The reader who doesn't see this has simply not learned to read.[48]

Do you see the problem here? It's only in the modern era that we've consciously distinguished these literary genres from each other. This allows us to write fictional stories in a style that reads like an account of real events. In a sense, that's what novels are. It's not as if novel writers are trying to trick anyone into thinking the events actually happened; it's a tool for telling a story, and the audience is fully aware of it. And novel writing like this is a product of the modern age, although there were forerunners to it. But to say that the Gospels are nonhistorical accounts written as if they were Graeco-Roman biographies would require us to believe that a group of first-century people foresaw modern literary analysis—despite there being no development of the idea prior to or after them—kept this knowledge a secret and, with no apparent motive, used it to construct a false set of stories that didn't happen but would read as

---

46. "Modern Theology and Biblical Criticism," in C. S. Lewis, *Christian Reflections*, 155.

47. Burridge, *What Are the Gospels?* In the foreword to the second edition, Graham Stanton writes that the first edition of this book "turned the tide" of scholarly assessment of the gospels' literary genre and that "I do not think that it is now possible to deny that the Gospels are a sub-set of the broad ancient literary genre of 'lives,' that is, biographies" (viii–ix).

48. "Modern Theology and Biblical Criticism," in C. S. Lewis, *Christian Reflections*, 155.

if they did. At this point, we've gone so far into conspiracy theory territory we might just spontaneously combust. It's about on the same level as saying that Shakespeare's plays were *really* social commentaries on the Trump presidencies.

As Lewis's last quote makes clear, this does not imply that the Gospels and Acts are holy writ without error. But it *does* imply that the central claims are roughly correct because those central claims are precisely what the stories are about. If we deny that those claims are roughly correct, we're beginning to move into the above territory where we ascribe the discovery of modern literary analysis to people from the first century.

That leads to the issue of miracles. There are plenty of historical texts that contain miraculous or supernatural elements, but they are usually later developments and are on the fringe of the story. They are not central claims but, rather, are adornments of the central claims, which allows us to dismiss them as unhistorical without having to enter into the whole discussion of the possibility of miracles. *But this option is not available for the Gospels* for a very obvious reason: the miraculous and supernatural elements *are* the central claims. You take them away and there's no story left—not even a "man travels around doing good deeds" story, because the good deeds Jesus performed were healings and other miracles.

So we have three reasons why we tend to dismiss accounts of miracles in history: first, they are later developments; second, they are not central to the story but are fringe elements; and third, because they're miracles and miracles can't happen. But when we try to apply this to Jesus' resurrection, we find that the first and second reasons don't apply. The resurrection is the earliest Christian claim, and it's not a fringe element of the story but its very center. That leaves us with the "miracles can't happen" argument, which I addressed in chapter 13. Most historical Jesus scholars reject the historicity of the resurrection for this third reason, that miracles just can't happen. But this claim is in the province of philosophy, not history, and the scholars in question are not philosophers but historians. So when they denounce miracles from the outset, they are speaking outside their area of authority.

V. RETURNING TO JESUS

I think that for many Christians (especially those who were raised Christian) when they encounter historical Jesus scholarship, they are horrified

The Christ Myth Myth

and scandalized by what the scholars deny. I came at it from the other direction: I grew up believing the Christ Myth was true, so when I encountered historical Jesus scholarship, I was horrified and scandalized by what the scholars *affirm*.

One claim the Christ Myth makes is that the idea that Jesus is God incarnate was not a part of the original Christian movement but emerged centuries later after a long period of mythologization. But a majority of biblical scholars today agree that the historical Jesus did see the one God of Israel as uniquely acting and speaking through him—in effect that his words and deeds were God's words and deeds. And his followers very quickly came to consider him to be the unique manifestation of Israel's God in the flesh: this idea was present from the very beginning, and in fact was the source of the whole Christian movement.[49] It's true it took a couple centuries to make *sense* of it: How could Jesus be the one God to whom he prayed? How could the ground of existence become a human being at all? Eventually, the church developed doctrines such as the Trinity (that God is three persons and one essence, Jesus being one of the persons) and the hypostatic union (that Jesus is fully God and fully human, not half and half) to explain it, but the idea that Jesus is somehow God incarnate was present at the beginning of Christianity, it was not something that only arose later.

To be sure, most scholars deny that Jesus directly *said* he was God (for that matter, most of them deny that he was God regardless). Most such statements, though not all, come in the Gospel of John. But scholars do acknowledge that Jesus *portrayed* himself as God by doing things that, in Judaism, only God could do, like forgiving people's sins against others. So Jesus—the historical Jesus, the actual guy who walked on the roads of Capernaum and Jerusalem—presented himself as the creator and sustainer of the universe. And again, most scholars do not believe he *was* the creator and sustainer of the universe, they just say that we can historically demonstrate that he portrayed himself as such.

This leads many people to ask, why should anyone believe him? Some answer by pointing to the ancient Christian argument, *aut Deus aut homo malus*—"either God or a bad man." Someone who portrays himself as the creator and sustainer of the universe is either desperately evil, desperately insane, or telling the truth; and as radical as that last option is, many people think it's more plausibly applied to Jesus than the others.[50]

49. Pöhlmann, *Abriss der Dogmatik*, 230.
50. For a contemporary defense of this argument and critique of the loopholes to it, see Kreeft, *Between Heaven and Hell*, which is a fictional debate—*obviously*

But another answer is the resurrection. Jesus was arrested by the religious leaders and executed by the political leaders for portraying himself as God and king. As stated earlier, for *that* guy to rise from the dead would amount to his divine vindication.[51]

So let's go over those claims again surrounding the resurrection that I went over in chapter 13. At the time, I presented them as alleged, but interestingly, as with the idea that Jesus portrayed himself as God, these claims are accepted as demonstrably historical by the consensus of scholars—most of whom do not believe miracles can happen and so don't believe that Jesus bodily rose from the dead. By "scholars" I mean "scholars in the relevant fields." If a scholar of pharmacological medicine disputed these claims, I wouldn't really care: we don't look to physicists to tell us about political science (or at least we *shouldn't*). And by "consensus" I mean the large majority of those scholars who publish on these issues in peer-reviewed journals and publishers. I'm not suggesting that someone's taken a poll of New Testament historians. And even if they had, I'd be more impressed by the former criterion: it's easy to say you don't believe something; it's much harder to publish specific objections in a peer-reviewed forum.

Note that I'm only *presenting* the claims of authorities; I'm not using them to argue that you should agree with them. Scholars and authorities can be wrong, after all. I'm not going over the actual reasons these scholars give for their conclusions because I'm not trying to argue that their conclusions are correct; just that these are, in fact, their conclusions. This is largely because my point here is not to defend the historicity of Jesus' resurrection. I am, first, distinguishing it from mythology. Second, I'm showing that *even if* there were parallels, the historical issues take priority. The most that mythological parallels could ever do is make you take a second look at the historical issues to make sure you didn't miss something. But since the alleged parallels can be found in actual historical accounts, their presence doesn't automatically mean an account is not historical.

So, again, I'm not trying to defend the historicity of Jesus' resurrection by listing the following contentions of contemporary scholarship. Granted, if I *were*, I'd probably go over the same claims, but I can't help that.

1. Jesus was brutally tortured and then executed by crucifixion.

---

fictional—between C. S. Lewis, John F. Kennedy, and Aldous Huxley in the moments after their deaths (they all died on November 22, 1963).

51. Pannenberg, *Glaube und Wirklichkeit*, 92–94.

2. Jesus' corpse was interred in a solid rock tomb that had a large boulder lodged in its entrance to seal it.

3. Jesus' tomb was discovered empty on the Sunday morning following his crucifixion.

4. Individuals and groups of up to five hundred people at once experienced appearances of Jesus alive from the dead. Some of them were not followers of Jesus, and some had actively opposed him.

5. The original Christian belief about this is that Jesus literally, bodily, physically rose from the dead. This belief arose immediately after it supposedly happened in the place where it supposedly happened (Jerusalem).

So, the consensus of relevant scholars—most of whom do not believe the resurrection took place because of that devastating "miracles can't happen" argument—agree that these claims can be historically demonstrated. Most of them—including, incredibly, the appearances of Jesus to hundreds of people after his death—are nearly universally accepted among the relevant scholars.[52] Moreover, none of these claims are intrinsically miraculous individually, they become so only when we try to formulate an explanatory theory to account for some or all of them together. Jesus' resurrection appearances, e.g. (point 4), suggest the miraculous only when combined with the fact that he had died a few days earlier (point 1); but people interacting with someone they know is not miraculous by itself.

There have been attempts to find an alternative, naturalistic explanation of these claims (like the evil twin theory),[53] but most scholars have rejected them. I'm not going to go over their reasons for doing so, just as I'm not going over their reasons for accepting the five claims surrounding Jesus' alleged resurrection. If you want to research it, go ahead,[54] but if you want to ignore it and never think about it again, go ahead. You do you. Godspeed.

---

52. Habermas, "Resurrection Research from 1975 to the Present." This is updated in Habermas, *Risen Indeed*, 4–28.

53. See above, 200.

54. As a starting point I'd suggest the debates of William Lane Craig. Craig defends the historicity of the resurrection with many interlocutors so it gives you both sides, along with plenty of further references. I originally got the five claims about the historical Jesus from Craig's debate with John Dominic Crossan, published as Copan, *Will the Real Jesus Please Stand Up?*

Bibliography

Aardsma, Gerald A. "Geocentricity and Creation." *Acts and Facts* 23 (1994). http://www.icr.org/article/geocentricity-creation.

Adams, Marilyn McCord. *Horrendous Evils and the Goodness of God*. Ithaca, NY: Cornell University Press, 2000.

Adams, Robert Merrihew. *Finite and Infinite Goods: A Framework for Ethics*. New York: Oxford University Press, 1999.

Aiton, E. J. "Galileo's Theory of the Tides." *Annals of Science* 10 (1954) 44–57.

Al-Khalili, Jim. "Science: Islam's Forgotten Geniuses." *Telegraph*, January 29, 2008. https://www.telegraph.co.uk/news/science/science-news/3323462/Science-Islams-forgotten-geniuses.html.

Alston, William. "What Euthyphro Should Have Said." In *Philosophy of Religion: A Reader and Guide*, edited by William Lane Craig, 283–98. New Brunswick, NJ: Rutgers University Press, 2001.

Amesbury, Richard. "Fideism." *Stanford Encyclopedia of Philosophy*, Summer 2022. Edited by Edward N. Zalta. https://plato.stanford.edu/archives/sum2022/entries/fideism/.

Amundsen, Darrel W. "Medieval Canon Law on Medical and Surgical Practice by the Clergy." *Bulletin of the History of Medicine* 52 (1978) 22–44.

Anselm. *Proslogion*. In *St. Anselm: Basic Writings*, translated by S. N. Deane, 1–34. 2nd ed. LaSalle, IL: Open Court, 1962.

Aquinas. *Summa Theologica*. In *Thomas Aquinas*, translated by Fathers of the English Dominican Province, vol. 1. Great Books of the Western World 19. Chicago: Encyclopædia Britannica, 1952.

Archer, Gleason. *Encyclopedia of Bible Difficulties*. Grand Rapids: Zondervan, 1982.

Aristotle. *On the Heavens (De Caelo)*. Translated by J. L. Stocks. In *The Works of Aristotle*, 1:359–405. Great Books of the Western World 8. Chicago: Encyclopædia Britannica, 1952.

———. *Politics (Politica)*. Translated by translated by Benjamin Jowett. In *The Works of Aristotle*, 2:445–548. Great Books of the Western World 9. Chicago: Encyclopædia Britannica, 1952.

Augustine. *The City of God*. Translated by Marcus Dods. In *Nicene and Post-Nicene Fathers*, 1st ser., edited by Philip Schaff, 2:1–511. 1887. Repr., Peabody, MA: Hendrickson, 1995.

———. *The Literal Meaning of Genesis*. Translated by John Hamond Taylor. 2 vols. Ancient Christian Writers: The Works of the Fathers in Translation 41. New York: Newman, 1982.

Avramides, Anita. "Other Minds." *Stanford Encyclopedia of Philosophy*, Winter 2023. Edited by Edward N. Zalta. https://plato.stanford.edu/archives/win2023/entries/other-minds/.

Baldwin, Erik, and Tyler Dalton McNabb. *Plantingian Religious Epistemology and World Religions: Prospects and Problems*. Studies in Comparative Philosophy and Religion. Lanham, MD: Lexington, 2019.

Baretti, Giuseppe. *The Italian Library: Containing an Account of the Lives and Works of the Most Valuable Authors of Italy*. London: Millar, 1757.

Barrow, John D. *The Book of Nothing: Vacuums, Voids, and the Latest Ideas About the Origins of the Universe*. New York: Pantheon, 2000.

Barrow, John D., and Frank J. Tipler. *The Anthropic Cosmological Principle*. New York: Oxford University Press, 1986.

Baur, C. "Severian." New Advent, 1912. From *The Catholic Encyclopedia*, vol. 13. http://www.newadvent.org/cathen/13742b.htm.

———. "Theodore of Mopsuestia." New Advent, 1912. From *The Catholic Encyclopedia*, vol. 14. http://www.newadvent.org/cathen/14571b.htm.

Bayrakdar, Mehmet. "Al-Jahiz and the Rise of Biological Evolution." *Islamic Quarterly* 27 (1983) 149–55.

Beatrice, Pier Franco. *The Transmission of Sin: Augustine and the Pre-Augustinian Sources*. Translated by Adam Kamesar. New York: Oxford University Press, 2013.

Benedict XVI. "Faith, Reason and the University: Memories and Reflections." Vatican, September 12, 2006. https://www.vatican.va/content/benedict-xvi/en/speeches/2006/september/documents/hf_ben-xvi_spe_20060912_university-regensburg.html.

Bercé, Yves-Marie, and Jean-Claude Otteni. "Pratique de la vaccination antivariolique dans les provinces de l'état pontifical au 19e s. Remarques sur le suppose interdit vaccinal de Léon XII." *Revue d'Histoire Ecclésiastique* 103 (2008) 448–66.

Bergson, Henri. *Creative Evolution*. Translated by Arthur Mitchell. New York: Holt, 1911.

Bevan, Edwyn R. "Mystery Religions and Christianity." In *Contemporary Thinking About Paul: An Anthology*, edited by Thomas S. Kepler, 42–48. New York: Abingdon-Cokesbury, 1940.

Blish, James. *A Case of Conscience*. New York: Ballantine, 1958.

Boethius. *The Consolation of Philosophy*. Translated by "I. T." and H. F. Stewart. In *Boethius*, 130–435. Loeb Classical Library 74. Cambridge, MA: Harvard University Press, 1973.

Bohm, David, and Basil J. Hiley. *The Undivided Universe: An Ontological Interpretation of Quantum Theory*. London: Routledge, 1995.

Boorstin, Daniel J. *The Discoverers: A History of Man's Search to Know His World and Himself*. New York: Random House, 1983.

Boylston, Arthur. "The Origins of Inoculation." *Journal of the Royal Society of Medicine* 105 (2012) 309–13.

Brague, Rémi. "Geocentrism as a Humiliation for Man." *Medieval Encounters* 3 (1997) 187–210.

Bresher, M. R. *The Newtonian System of Astronomy: With a Reply to the Various Objections Made Against It by "Parallax."* London: Whittaker, 1868.
Bristow, William. "Enlightenment." *Stanford Encyclopedia of Philosophy*, Fall 2023. Edited by Edward N. Zalta. https://plato.stanford.edu/archives/fall2023/entries/enlightenment/.
Brooke, John Hedley. "The Wilberforce-Huxley Debate: Why Did It Happen?" *Science and Christian Belief* 13 (2001) 127–41.
Brown, Fredric. "Solipsist." In *From These Ashes: The Complete Short SF of Fredric Brown*, 563. Framingham, MA: NESFA, 2000.
Browne, Thomas. *Religio Medici*. 1643. Repr., London: Low, Son, and Marsten, 1869.
Bruno, Giordano. *On the Infinite Universe and Worlds*. 1584. In *Giordano Bruno: His Life and Thought*, translated by Dorothea Waley Singer, 225–378. New York: Greenwood, 1968.
Burridge, Richard. *What Are the Gospels? A Comparison with Graeco-Roman Biography*. 1st ed. Cambridge: Cambridge University Press, 1992. 2nd ed. Grand Rapids: Eerdmans, 2004.
Burt, Donald X. "Augustine and Divine Voluntarism." *Angelicum* 64 (1987) 424–36.
Cairns-Smith, A. G. *Seven Clues to the Origin of Life: A Scientific Detective Story*. Cambridge: Cambridge University Press, 1990.
Campbell, George Douglas (Duke of Argyll). "What Is Science?" In *Good Words for 1885*, edited by Donald MacLeod, 26:244. London: Isbester, 1885.
———. *What Is Science?* Edinburgh: Douglas, 1898.
Campbell, Joseph. *The Hero with a Thousand Faces*. New York: Pantheon, 1949.
———. *Occidental Mythology*. Vol. 3 of *The Masks of God*. New York: Penguin, 1964.
Camus, Albert. *The Plague*. Translated by Stuart Gilbert. New York: Knopf Doubleday, 1991.
Carile, Maria Cristina. "Globus Cruciger." In *Encyclopedia of the Bible and Its Reception*, edited by Constance M. Furey et al., 10:299–301. Berlin: de Gruyter, 2015.
Carré, Lorenzo, et al. "Relevance of Earth-Bound Extremophiles in the Search for Extraterrestrial Life." *Astrobiology* 22 (2022) 322–67.
Cavin, Robert Greg. "Miracles, Probability, and the Resurrection of Jesus: A Philosophical, Mathematical, and Historical Study." PhD diss., University of California, Irvine, 1993.
Cawte, E. C. "It's an Ancient Custom—but How Ancient?" In *Aspects of British Calendar Customs*, edited by Theresa Buckland and Juliette Wood, 37–56. Sheffield: Sheffield Academic Press, 1993.
Chalmers, Robert. *Vestiges of the Natural History of Creation*. London: Churchill, 1844.
Chapman, John. "Council of Ephesus." New Advent, 1909. From *The Catholic Encyclopedia*, vol. 5. http://www.newadvent.org/cathen/05491a.htm.
———. "St. Cyril of Alexandria." New Advent, 1908. From *The Catholic Encyclopedia*, vol. 4. http://www.newadvent.org/cathen/04592b.htm.
Chela-Flores, Julian. "The Phenomenon of the Eukaryotic Cell." In *Evolutionary and Molecular Biology: Scientific Perspectives on Divine Action*, edited by Robert Russell et al., 79–98. Vatican: Vatican Observatory and Center for Theology and the Natural Sciences, 1998.
Chiang, Ted. "Omphalos." In *Exhalation*, 237–69. New York: Vintage, 2019.

Chrysostom, John. *Homilies of St. John Chrysostom on the Epistle to the Hebrews.* Oxford translation. In *Nicene and Post-Nicene Fathers,* 1st ser., edited by Philip Schaff, 14:335–522. 1889. Repr., Peabody, MA: Hendrickson, 1995.

Churchland, Patricia. *Neurophilosophy: Toward a Unified Science of the Mind-Brain.* Cambridge, MA: MIT Press, 1986.

Clarke, Arthur C. "The Wall of Darkness." In *The Collected Stories of Arthur C. Clarke,* 104–18. New York: Doherty, 1958.

Collins, Robin. "The Teleological Argument: An Exploration of the Fine-Tuning of the Universe." In *The Blackwell Companion to Natural Theology,* edited by William Lane Craig and J. P. Moreland, 202–81. Malden, MA: Wiley-Blackwell, 2009.

Condé, Maryse. *Crossing the Mangrove.* New York: Random House, 1995.

Cook, Harold J. *Matters of Exchange: Commerce, Medicine, and Science in the Dutch Golden Age.* New Haven, CT: Yale University Press, 2007.

Copan, Paul, ed. *Will the Real Jesus Please Stand Up? A Debate Between William Lane Craig and John Dominic Crossan.* Moderated by William F. Buckley Jr. Grand Rapids: Baker, 1998.

Copan, Paul, and William Lane Craig, eds. *Come Let Us Reason: New Essays in Chrisian Apologetics.* Nashville: B&H, 2012.

Copenhaver, Brian P. "Introduction." In *Hermetica: The Greek "Corpus Hermeticum" and the Latin "Asclepius" in a New English Translation, with Notes and Introduction,* translated by Brian P. Copenhaver, xiii–lxi. Cambridge: Cambridge University Press, 1992.

Copernicus, Nicolaus. *On the Revolutions of the Heavenly Spheres.* 1543. Translated by Charles Glenn Wallis. In *Ptolemy Copernicus Kepler,* 497–838. Great Books of the Western World 16. Chicago: Encyclopædia Britannica, 1952.

Copleston, Frederick. *A History of Philosophy.* 9 vols. Westminster, MD: Newman, 1946–75.

Corrigan, Kevin, and L. Michael Harrington. "Pseudo-Dionysius the Areopagite." *Stanford Encyclopedia of Philosophy,* Summer 2023. Edited by Edward N. Zalta. https://plato.stanford.edu/archives/sum2023/entries/pseudo-dionysius-areopagite/.

Craig, William Lane. *Reasonable Faith: Christian Truth and Apologetics.* 3rd ed. Wheaton, IL: Crossway, 2008.

Crowe, Michael. *The Extraterrestrial Life Debate, 1750–1900: The Idea of a Plurality of Worlds from Kant to Lowell.* Cambridge: Cambridge University Press, 1986.

Custance, Arthur C. *Without Form and Void: A Study of the Meaning of Genesis 1:2.* Brockville, Can.: Doorway, 1970.

D'Evelyn, Charlotte, and Anna J. Mill, eds. *Text.* Vol. 2 of *The South England Legendary.* Early English Text Society. New York: Oxford University Press, 1967.

Dandelet, Thomas James. *Spanish Rome, 1500–1700.* New Haven, CT: Yale University Press, 2001.

Danielson, Dennis. "Copernicus and the Tale of the Pale Blue Dot." Lecture presented at Pascal Lectures, University of Waterloo, Ontario, Canada, October 20, 2004.

———. *The First Copernican: Georg Joachim Rheticus and the Rise of the Copernican Revolution.* New York: Walker, 2006.

———. "The Great Copernican Cliché." *American Journal of Physics* 69 (2001) 1029–35.

Dantzig, Tobias. *Number: The Language of Science: A Critical Survey Written for the Cultured Non-Mathematician*. New York: Macmillan, 1930.

Darwin, Charles. *The Autobiography of Charles Darwin, 1809–1882, with Original Omissions Restored*. Edited by Nora Barlow. London: Collins, 1958.

———. *The Life and Letters of Charles Darwin, Including an Autobiographical Chapter*. Edited by Francis Darwin. Vol. 1. London: Murray, 1887.

———. *On the Origin of Species by Means of Natural Selection, or the Preservation of Favoured Races in the Struggle for Life*. 6th ed. London: Murray, 1873.

Dennett, Daniel C. *Darwin's Dangerous Idea: Evolution and the Meanings of Life*. New York: Simon & Schuster, 1995.

———. *Intuition Pumps and Other Tools for Thinking*. New York: Norton, 2013.

Dennett, Daniel C., and Alvin Plantinga. *Science and Religion: Are They Compatible?* New York: Oxford University Press, 2011.

Dick, Steven J. *Plurality of Worlds: The Origins of the Extraterrestrial Life Debate from Democritus to Kant*. Cambridge: Cambridge University Press, 1984.

Dilman, İlham. *Philosophy as Criticism: Essays on Dennett, Searle, Foot, Davidson, Nozick*. New York: Continuum, 2011.

Doig, James. "Aquinas and Aristotle." In *The Oxford Handbook of Aquinas*, edited by Brian Davies and Eleonore Stump, 33–44. Oxford Handbooks. New York: Oxford University Press, 2012.

Downie, R. Angus. *Frazer and the Golden Bough*. London: Gallancz, 1970.

Doyle, Arthur Conan. *The Sign of Four*. London: Blackett, 1890.

Draper, John William. *History of the Conflict Between Science and Religion*. New York: Appleton, 1874.

———. *A History of the Intellectual Development of Europe*. Rev. ed. 2 vols. London: Bell and Sons, 1875.

Drews, Carl, and Weiqing Han. "Dynamics of Wind Setdown at Suez and the Eastern Nile Delta." *PLOS ONE* 5 (2010). https://opensky.ucar.edu/islandora/object/articles:10312/datastream/PDF/view.

Duhem, Pierre. *Le système du monde: Histoire des doctrines cosmologiques de Platon à Copernic*. 10 vols. Paris: Hermann, 1914.

Dundes, Alan, ed. *The Flood Myth*. Berkeley: University of California Press, 1988.

Dunn, James D. G. *Beginning from Jerusalem*. Vol. 2 of *Christianity in the Making*. Grand Rapids: Eerdmans, 2009.

———. *Romans 1–8*. Word Biblical Commentary 38a. Grand Rapids: Zondervan, 1988.

Durbach, Nadja. "'They Might as Well Brand Us': Working-Class Resistance to Compulsory Vaccination in Victorian England." *Social History of Medicine* 13 (2000) 45–63.

Dzielska, Maria. *Hypatia of Alexandria*. Translated by F. Lyra. Cambridge, MA: Harvard University Press, 1996.

Earman, John. *Hume's Abject Failure: The Argument Against Miracles*. New York: Oxford University Press, 2000.

Edwards, Mark J. "Origen." *Stanford Encyclopedia of Philosophy*, Summer 2022. Edited by Edward N. Zalta. https://plato.stanford.edu/archives/sum2022/entries/origen/.

Egerton, Frank N. "A History of the Ecological Sciences, Part 6: Arabic Language Science: Origins and Zoological Writings." *Bulletin of the Ecological Society of America* 83 (2002) 142–46.

England, Richard. "Censoring Huxley and Wilberforce: A New Source for the Meeting That the *Athenaeum* 'Wisely Softened Down.'" *Notes and Records: The Royal Society Journal of the History of Science* 71 (2017) 371–84.

Erickson, Millard J. *Christian Theology*. 2nd ed. Grand Rapids: Baker, 1998.

Everitt, Nicholas. *The Non-Existence of God*. London: Routledge, 2004.

Fakhry, Majid. *A History of Islamic Philosophy*. 2nd ed. New York: Columbia University Press, 1983.

Falcon, Andrea. "Aristotle on Causality." *Stanford Encyclopedia of Philosophy*, Spring 2023. Edited by Edward N. Zalta. https://plato.stanford.edu/archives/spr2023/entries/aristotle-causality/.

Fales, Evan. "Successful Defense? A Review of *In Defense of Miracles*." *Philosophia Christi* 3 (2001) 7–35.

Farley, John. *The Spontaneous Generation Controversy from Descartes to Oparin*. Baltimore: Johns Hopkins University Press, 1977.

Farr, A. D. "Early Opposition to Obstetric Anesthesia." *Anesthesia* 35 (1980) 896–907.

———. "Religious Opposition to Obstetric Anesthesia: A Myth?" *Annals of Science* 40 (1983) 159–77.

Ferngren, Gary B. *Medicine and Religion: A Historical Introduction*. Baltimore: Johns Hopkins University Press, 2014.

Finocchiaro, Maurice A. *The Galileo Affair: A Documentary History*. Berkeley: University of California Press, 1989.

———, ed. and trans. *The Trial of Galileo: Essential Documents*. Indianapolis: Hackett, 2014.

Flynn, Michael. "The Common Goal of Nature." In *"The Forest of Time" and Other Stories*. New York: Tor, 1997.

Fontenrose, Joseph Eddy. *The Ritual Theory of Myth*. Berkeley: University of California Press, 1966.

Frankfurt, Harry G. *On Bullshit*. Princeton, NJ: Princeton University Press, 2005.

Frazer, James George. *The Golden Bough: A Study in Magic and Religion*. 13 vols. Repr., New York: Macmillan, 1976.

Freud, Sigmund. "A Difficulty in the Path of Psychoanalysis." Translated by Joan Riviere. In *Psychological Writings and Letters*, edited by Sander L. Gilman. New York: Continuum, 1995.

Friedman, Maurice. *Martin Buber's Life and Work: The Early Years, 1878–1923*. New York: Dutton, 1981.

Galilei, Galileo. *Dialogue Concerning the Two Chief World Systems*. Translated by Thomas Salusbury. London: Leybourne, 1656.

———. "Letter to the Grand Duchess Christina." In *Discoveries and Opinions of Galileo*, translated by Stillman Drake, 173–216. New York: Random House, 1957.

———. *Sidereus Nuncius, or The Sidereal Messenger*. Translated by Albert Van Helden. Chicago: University of Chicago Press, 1989.

Garfinkle, Richard. *Celestial Matters*. New York: Tor, 1996.

Garwood, Christine. *Flat Earth: The History of an Infamous Idea*. London: Macmillan, 2007.

Gerard, John. "Galileo Galilei." New Advent, 1909. From *The Catholic Encyclopedia*, vol. 6. http://www.newadvent.org/cathen/06342b.htm.

Godkin, G. S. *Life of Victor Emmanuel II, First King of Italy*. Vol. 1. London: Macmillan, 1880.

Goldie, Matthew Boyd. *The Idea of the Antipodes: Place, People, and Voices*. Routledge Research in Postcolonial Literatures. London: Routledge, 2010.

Goldstein, Bernard R. "The Arabic Version of Ptolemy's Planetary Hypotheses." *Transactions of the American Philosophical Society*, n.s., 57 (1967) 3–55.

González, Justo L. "Athens and Jerusalem Revisited: Reason and Authority in Tertullian." *Church History* 43 (1974) 17–25.

Gould, Stephen Jay. *The Mismeasure of Man*. New York: Norton, 1981.

Graney, Christopher M. *Setting Aside All Authority: Giovanni Battista Riccioli and the Science Against Copernicus in the Age of Galileo*. Notre Dame, IN: University of Notre Dame Press, 2015.

Habermas, Gary. "Resurrection Research from 1975 to the Present: What Are Critical Scholars Saying?" *Journal for the Study of the Historical Jesus* 3 (2005) 135–53.

———. *Risen Indeed: A Historical Investigation into the Resurrection of Jesus*. Bellingham, WA: Lexham Academic, 2021.

Habermas, Gary R., and J. P. Moreland. *Immortality: The Other Side of Death*. Nashville: Thomas Nelson, 1992.

Haddon, A. C. *Magic and Fetishism*. Cambridge: Cambridge University Press, 1910.

Hagopian, David G., ed. *The Genesis Debate: Three Views on the Days of Creation*. Mission Viejo, CA: Crux, 2001.

Hankinson, R. J. "The Man and His Work." In *The Cambridge Companion to Galen*, edited by R. J. Hankinson, 1–33. Cambridge: Cambridge University Press, 2008.

Hannam, James. *The Globe: How the Earth Became Round*. London: Reaktion, 2023.

———. *God's Philosophers: How the Medieval World Laid the Foundations of Modern Science*. London: Icon, 2009. Also published as *The Genesis of Science: How the Christian Middle Ages Launched the Scientific Revolution*. Washington DC: Regnery, 2011.

Harent, Stéphane. "Original Sin." New Advent, 1911. From *The Catholic Encyclopedia*, vol. 11. http://www.newadvent.org/cathen/11312a.htm.

Harrison, Edward Robert. *Darkness at Night: A Riddle of the Universe*. Cambridge, MA: Harvard University Press, 1987.

Harrison, Peter. *The Bible, Protestantism, and the Rise of Natural Science*. Cambridge: Cambridge University Press, 1998.

———. "The Development of the Concept of Laws of Nature." In *Creation: Law and Probability*, edited by Fraser Watts, 13–36. Aldershot, UK: Ashgate, 2007.

———. *The Fall of Man and the Foundations of Science*. Cambridge: Cambridge University Press, 2008.

Hawking, Stephen. *A Brief History of Time: From the Big Bang to Black Holes*. New York: Bantam, 1988.

Hawking, Stephen, and Leonard Mlodinow. *The Grand Design*. New York: Bantam 2010.

Hegel, George Wilhelm Friedrich. *Lectures on the History of Philosophy*. Translated by E. S. Haldane. Vol. 3. London: Kegan Paul, Trench, Trübner & Co, 1892.

Helmig, Christoph, and Carlos Steel. "Proclus." *Stanford Encyclopedia of Philosophy*, Fall 2021. Edited by Edward N. Zalta. https://plato.stanford.edu/archives/fall2021/entries/proclus/.

Henderson, Leah. "The Problem of Induction." *Stanford Encyclopedia of Philosophy*, Winter 2022. Edited by Edward N. Zalta. https://plato.stanford.edu/archives/win2022/entries/induction-problem/.

Herrick, Paul. *Introduction to Logic*. New York: Oxford University Press, 2013.

———. *Philosophy, Reasoned Belief, and Faith: An Introduction*. Notre Dame, IN: University of Notre Dame Press, 2022.

———. *Think with Socrates: An Introduction to Critical Thinking*. New York: Oxford University Press, 2015.

Hess, Peter M. J. "Two Books." In *Encyclopedia of Science and Religion*, edited by J. Wentzel Vrede van Huyssteen, 2:905–8. New York: Macmillan Reference, 2003. https://www.encyclopedia.com/education/encyclopedias-almanacs-transcripts-and-maps/two-books.

Hilgevoord, Jan, and Jos Uffink. "The Uncertainty Principle." *Stanford Encyclopedia of Philosophy*, Spring 2024. Edited by Edward N. Zalta. https://plato.stanford.edu/archives/spr2024/entries/qt-uncertainty/.

Hooke, S. H. "Christianity and the Mystery Religions." In *Judaism and Christianity*, edited by W. O. E. Oesterly, 1:235–50. New York: Macmillan, 1937.

Howard-Snyder, Daniel, and Adam Green. "Hiddenness of God." *Stanford Encyclopedia of Philosophy*, Fall 2022. Edited by Edward N. Zalta. https://plato.stanford.edu/archives/fall2022/entries/divine-hiddenness/.

Hoyle, Fred, and Chandra Wickramasinghe. *Evolution from Space*. New York: Simon and Schuster, 1981.

Hume, David. *Enquiry Concerning Human Understanding*. 1748. Early Modern Texts, 2017. Edited by Jonathan Bennett. https://earlymoderntexts.com/assets/pdfs/hume1748.pdf.

———. *Treatise of Human Nature*. 1739. Early Modern Texts, 2017. Edited by Jonathan Bennett. https://earlymoderntexts.com/assets/pdfs/hume1739book1.pdf.

Humphreys, D. Russell. "Our Galaxy Is the Centre of the Universe, 'Quantized' Redshifts Show." *Journal of Creation (TJ)* 16 (2002) 102–3. https://creation.com/our-galaxy-is-the-centre-of-the-universe-quantized-redshifts-show.

Hunter, George William. *Civic Biology: Presented in Problems*. New York: American, 1914.

Irving, Washington. *A History of the Life and Voyages of Christopher Columbus*. 4 vols. London: Murray, 1828.

Ivry, Alfred L. "Averroes' Three Commentaries on *De anima*." In *Averroes and the Aristotelian Tradition: Sources, Constitution and Reception of the Philosophy of Ibn Rushd (1126–1198); Proceedings of the Fourth Symposium Averroicum (Cologne, 1996)*, edited by Gerhard Endress and Jan A. Aertsen, 199–216. Islamic Philosophy, Theology and Science. Texts and Studies 31. Leiden: Brill, 1999.

Jaki, Stanley L. *The Road of Science and the Ways to God*. Chicago: University of Chicago Press, 1978.

Janssens, Louis. "Artificial Insemination: Ethical Considerations." *Louvain Studies* 8 (1980) 3–29.

Jeannière, A. "Corps malleable." *Cahiers Laënnec* 29 (1968) 92–102.

Jensen, J. Vernon. "Return to the Wilberforce-Huxley Debate." *British Journal for the History of Science* 21 (1988) 161–79.

Jouanna, Jacques. "The Legacy of the Hippocratic Treatise *The Nature of Man*: The Theory of the Four Humours." In *Greek Medicine from Hippocrates to Galen: Selected Papers*, edited by Philip van der Eijk, translated by Neil Allies, 335–59. Studies in Ancient Medicine 40. Leiden: Brill, 2012.

Keating, Karl. *The New Geocentrists*. San Diego: Rasselas, 2015.

Keefe, Donald J. "Tracking a Footnote." *Fellowship of Catholic Scholars Newsletter* 9 (1986) 6–7.

King, Lester S., and Marjorie C. Meehan. "A History of the Autopsy: A Review." *American Journal of Pathology* 73 (1973) 514–44.

Knox, Dilwyn. "Giordano Bruno." *Stanford Encyclopedia of Philosophy*, Summer 2019. Edited by Edward N. Zalta. https://plato.stanford.edu/archives/sum2019/entries/bruno/.

Koestler, Arthur. *The Sleepwalkers: A History of Man's Changing Vision of the Universe*. London: Hutchinson, 1959.

Koritansky, Peter Karl. "Thomas Aquinas and the Euthyphro Dilemma." *Heythrop Journal* 62 (2021) 1013–24.

Koyré, Alexandre. *From the Closed World to the Infinite Universe*. Baltimore: Johns Hopkins University Press, 1957.

Krauss, Lawrence M. "The End of the Age Problem, and the Case for a Cosmological Constant Revisited." *Astrophysical Journal* 501 (1998) 461–66.

Kreeft, Peter. *Between Heaven and Hell*. Downers Grover, IL: InterVarsity, 1982.

Kreeft, Peter, and Ronald Tacelli. *Handbook of Christian Apologetics*. Downers Grove, IL: InterVarsity, 1994.

Lapide, Pinchas. *The Resurrection of Jesus: A Jewish Perspective*. Translated by Wilhelm C. Linss. Minneapolis: Augsburg, 1983.

Laplace, Pierre-Simon. *Exposition du système du monde*. 6th ed. Vol. 2. Paris: Bachelier, 1836. *The System of the World*. Translated by J. Pond. Vol. 2. London: Philips, 1809.

Latourette, Kenneth Scott. *Beginnings to 1500*. Rev. ed. Vol. 1 of *A History of Christianity*. New York: HarperCollins, 1975.

Leftow, Brian. *God and Necessity*. New York: Oxford University Press, 2012.

Lewis, C. S. *The Abolition of Man*. New York: Macmillan, 1947.

———. *Christian Reflections*. Edited by Walter Hooper. Grand Rapids: Eerdmans, 1967.

———. *The Discarded Image: An Introduction to Medieval and Renaissance Literature*. Cambridge: Cambridge University Press, 1964.

———. *God in the Dock*. Edited by Walter Hooper. Grand Rapids: Eerdmans, 1970.

———. *Mere Christianity*. New York: Macmillan, 1960.

———. *Miracles: A Preliminary Study*. 2nd ed. Glasgow: Fontana, 1960.

———. *The Pilgrim's Regress: An Allegorical Apology for Christianity, Reason and Romanticism*. Grand Rapids: Eerdmans, 1981.

———. *The Problem of Pain*. New York: Macmillan, 1962.

———. *Surprised by Joy: The Shape of My Early Life*. London: Collins, 1959.

———. *Till We Have Faces*. San Diego: Harcourt Brace & Co., 1957.

———. *The Weight of Glory and Other Addresses*. San Francisco: HarperCollins, 2001.

———. *The World's Last Night and Other Essays*. New York: Harcourt Brace Jovanovich, 1960.

Lewis, Gordon R., and Bruce A. Demarest. *Integrative Theology*. Vol. 1. Grand Rapids: Zondervan, 1996.

Lewis, Jason E., et al. "The Mismeasure of Science: Stephen Jay Gould Versus Samuel George Morton on Skulls and Bias." *PLOS Biology* 9 (2011). https://journals.plos.org/plosbiology/article?id=10.1371/journal.pbio.1001071.

Lindberg, David C. *The Beginnings of Western Science: The European Scientific Tradition in Philosophical, Religious, and Institutional Context, 600 B.C. to A.D. 1450*. Chicago: University of Chicago Press, 1992.

———. "The Transmission of Greek and Arabic Learning to the West." In *Science in the Middle Ages*, edited by David C. Lindberg, 52–90. Chicago: University of Chicago Press, 1978.

Lindberg, David C., and Ronald L. Numbers. "Beyond War and Peace: A Reappraisal of the Encounter Between Christianity and Science." *Church History* 55 (1986) 338–54.

———, eds. *God and Nature: Historical Essays on the Encounter Between Christianity and Science*. Berkeley: University of California Press, 1986.

———, eds. *When Science and Christianity Meet*. Chicago: University of Chicago Press, 2003.

Liu, Cixin. *Death's End*. Translated by Ken Liu. New York: Tor, 2016.

Livingstone, David N. *Darwin's Forgotten Defenders: The Encounter Between Evangelical Theology and Evolutionary Thought*. Grand Rapids: Eerdmans, 1987.

Lovejoy, Arthur. *The Great Chain of Being: A Study of the History of an Idea*. Cambridge, MA: Harvard University Press, 1964.

Lucas, J. R. *The Freedom of the Will*. Oxford: Clarendon, 1970.

Lyons, Jonathan. *The House of Wisdom: How the Arabs Transformed Western Civilization*. London: Bloomsbury, 2009.

Machen, J. Gresham. *The Origin of Paul's Religion*. New York: Macmillan, 1921.

———. *The Virgin Birth of Christ*. New York: Harper & Row, 1930.

Macleod, Ken. "A Case of Consilience." In *The Year's Best Science Fiction: Twenty-Third Annual Collection*, edited by Gardner Dozois, 171–80. New York: St. Martin's, 2006.

Mann, Charles C. *1491: New Revelations of the Americas Before Columbus*. New York: Knopf, 2005.

Marrett, R. R., and T. K. Penniman, eds. *Spencer's Scientific Correspondence with Sir J. G. Frazer and Others*. Oxford: Clarendon, 1932.

Mayr, Ernst. "How to Carry Out the Adaptationist Program?" *American Naturalist* 121 (1983) 324–34.

McCartney, Eugene S. "Spontaneous Generation and Kindred Notions in Antiquity." *Transactions of the American Philological Association* 51 (1920) 101–15.

McColley, Grant, and H. W. Miller. "Saint Bonaventure, Francis Mayron, William Vorilong, and the Doctrine of a Plurality of Worlds." *Speculum* 12 (1937) 386–89.

McCormick, Richard. *Health and Medicine in Catholic Tradition: Tradition in Transition*. New York: Crossroad, 1984.

McDevitt, Jack. "Gus." In *Sacred Visions*, edited by Andrew M. Greely and Michael Cassutt, 1–25. New York: Tor, 1991.

McGrath, Alister E. *Iustitia Dei: A History of the Christian Doctrine of Justification*. 3rd ed. Cambridge: Cambridge University Press, 2005.

———. *Science and Religion: An Introduction*. Malden: Blackwell, 2000.

McGrew, Timothy. "Miracles." *Stanford Encyclopedia of Philosophy*, Spring 2024. Edited by Edward N. Zalta. https://plato.stanford.edu/archives/spr2024/entries/miracles/.

McMullin, Ernan. "The Church's Ban on Copernicanism, 1616." In *The Church and Galileo*, edited by Ernan McMullin, 150–90. Notre Dame, IN: University of Notre Dame Press, 2005.

Melander, Peter. *Analyzing Functions: An Essay on a Fundamental Notion in Biology*. Stockholm: Almqvist & Wiksell, 1997.

Metzger, Bruce M. "Methodology in the Study of the Mystery Religions and Early Christianity." In *Historical and Literary Studies: Pagan, Jewish, and Christian*, 1–24. Grand Rapids: Eerdmans, 1968.

Michaelson, Eliot. "Reference." *Stanford Encyclopedia of Philosophy*, Summer 2024. Edited by Edward N. Zalta. https://plato.stanford.edu/entries/reference/.

Michell, John. *Eccentric Lives and Peculiar Notions*. New York: Harcourt Brace Jovanovich, 1984.

Moore, James R. *The Post-Darwinian Controversies: A Study of the Protestant Struggle to Come to Terms with Darwin in Great Britain and America, 1870–1900*. Cambridge: Cambridge University Press, 1979.

Moran, Dermot, and Adrian Guiu. "John Scotus Eriugena." *Stanford Encyclopedia of Philosophy*, Winter 2021. Edited by Edward N. Zalta. https://plato.stanford.edu/archives/win2021/entries/scottus-eriugena/.

Morris, Thomas V. *The Logic of God Incarnate*. Ithaca, NY: Cornell University Press, 1986.

Morton, Samuel George. *Crania Americana: or, A Comparative View of the Skulls of Various Aboriginal Nations of North and South America*. Philadelphia: Dobson, 1839.

Munn, Norman L. *Introduction to Psychology*. Boston: Houghton Mifflin, 1962.

Murray, William J. *My Life Without God*. Eugene, OR: Harvest, 1992.

Murschel, Andrea. "The Structure and Function of Ptolemy's Physical Hypotheses of Planetary Motion." *Journal for the History of Astronomy* 26 (1995) 33–61.

Nagel, Thomas. *The Last Word*. New York: Oxford University Press, 1997.

———. *Mind and Cosmos: Why the Materialist Neo-Darwinian Conception of Nature Is Almost Certainly False*. New York: Oxford University Press, 2012.

———. *The View from Nowhere*. New York: Oxford University Press, 1986.

Nails, Debra, and S. Sara Monoson. "Socrates." *Stanford Encyclopedia of Philosophy*, Summer 2022. Edited by Edward N. Zalta. https://plato.stanford.edu/archives/sum2022/entries/socrates.

Nash, Ronald. *The Gospel and the Greeks: Did the New Testament Borrow from Pagan Thought?* 2nd ed. Phillipsburg, NJ: P&R, 2003.

Neugebauer, Otto. *The Exact Sciences in Antiquity*. New York: Harper & Bros., 1962.

Newman, John Henry. *The Letters and Diaries of John Henry Newman*. Edited by C. S. Dessain and T. Gornall. Vol. 24. Oxford: Clarendon, 1973.

Nietzsche, Friedrich. "On Truth and Lies in a Nonmoral Sense." Edited and translated by Daniel Breazeale. In *Truth: Engagements Across Philosophical Traditions*, edited by José Medina and David Wood, 14–25. Malden, MA: Blackwell, 2005.

Nissen, Lowell. *Teleological Language in the Life Sciences*. Lanham, MD: Rowman & Littlefield, 1997.

Numbers, Ronald L. *The Creationists: From Scientific Creationism to Intelligent Design*. Cambridge, MA: Harvard University Press, 2006.

———, ed. *Galileo Goes to Jail and Other Myths About Science and Religion*. Cambridge, MA: Harvard University Press, 2009.

O'Conner, Patricia T., and Stewart Kellerman. *Origins of the Specious: Myths and Misconceptions of the English Language*. New York: Random House, 2009.

O'Malley, C. D. *Andreas Vesalius of Brussels, 1514–1564*. Berkeley: University of California Press, 1964.

Origen. *De Principiis*. Edinburgh translation. In *Ante-Nicene Fathers*, edited by Alexander Roberts and James Donaldson, 4:239–384. 1885. Repr., Peabody, MA: Hendrickson, 1995.

Orr, James. "The Early Narratives of Genesis." In *The Fundamentals: A Testimony to the Truth*, edited by R. A. Torrey et al., 6:85–97. Chicago: Testimony, 1910.

———. "Science and Christian Faith." In *The Fundamentals: A Testimony to the Truth*, edited by R. A. Torrey et al., 4:91–104. Chicago: Testimony, 1910.

Packard, Alpheus S. *Lamarck, the Founder of Evolution: His Life and Work*. New York: Longmans, Green, and Co., 1901.

Paley, William. *Natural Theology or Evidences of the Existence and Attributes of the Deity*. London: Faulder, 1802.

Pannenberg, Wolfhart. *Glaube und Wirklichkeit*. Munich: Kaiser, 1975.

———. *Systematic Theology*. Translated by Geoffrey W. Bromiley. Vol. 2. Grand Rapids: Eerdmans, 1994.

Pascal, Blaise. *Pensées*. Edited and translated by Roger Ariew. Indianapolis: Hackett, 2005.

Paul VI. "*Humanae Vitae*: On the Regulation of Birth." Vatican, July 25, 1968. https://www.vatican.va/content/paul-vi/en/encyclicals/documents/hf_p-vi_enc_25071968_humanae-vitae.html.

Pellat, Charles. *The Life and Works of Jāḥiẓ: Translations of Selected Texts*. Translated by D. M. Hawke. Berkeley: University of California Press, 1969.

Penrose, Roger. *The Road to Reality: A Complete Guide to the Laws of the Universe*. New York: Knopf, 2005.

Peppiatt, Lucy. *The Imago Dei: Humanity Made in the Image of God*. Eugene, OR: Cascade, 2022.

Peters, Ted, et al., eds. *Astrotheology: Science and Theology Meet Extraterrestrial Life*. Eugene, OR: Cascade, 2018.

Peters, Ted, and Julie Froehlig. "The Peters ETI Religious Crisis Survey." Counterbalance, 2008. http://www.counterbalance.net/etsurv/PetersETISurveyRep.pdf.

Peterson, Michael L. *The Problem of Evil: Selected Readings*. 2nd ed. Notre Dame, IN: University of Notre Dame Press, 2017.

Philipse, Herman. *God in the Age of Science? A Critique of Religious Reason*. New York: Oxford University Press, 2012.

Phillpotts, B. "German Heathenism." In *The Rise of the Saracens and the Foundation of the Western Empire*, edited by J. B. Bury et al., 480–95. Vol. 2 of *The Cambridge Medieval History*. New York: Macmillan, 1913.

Plantinga, Alvin. *The Nature of Necessity*. New York: Oxford University Press, 1974.

———. *Warrant and Proper Function*. New York: Oxford University Press, 1993.

———. *Where the Conflict Really Lies: Science, Religion, and Naturalism*. New York: Oxford University Press, 2011.

Plantinga, Alvin, and Michael Tooley. *Knowledge of God*. Malden, MA: Blackwell, 2008.

Plutarch. "Isis and Osiris." In *Moralia*, translated by Frank Cole Babbitt, 5:6–191. Loeb Classical Library 306. Cambridge, MA: Harvard University Press, 1936.

Pöhlmann, Horst Georg. *Abriss der Dogmatik*. 3rd ed. Gütersloh: Gerd Mohn, 1980.

Pontin, Jason. "A Q&A with Gene Wolfe." *MIT Technology Review* 117 (2014). https://www.technologyreview.com/s/529431/a-qa-with-gene-wolfe.

Porter, Roy. *Blood and Guts: A Short History of Medicine*. New York: Norton, 2002.

Proctor, Richard A. *Myths and Marvels of Astronomy*. New York: Putnam's Sons, 1877.

Pruss, Alexander R., and Joshua L. Rasmussen. *Necessary Existence*. New York: Oxford University Press, 2018.

Ptolemy. *Almagest*. Translated by R. Catesby Talioferro. In *Ptolemy Copernicus Kepler*, 1–478. Great Books of the Western World 16. Chicago: Encyclopædia Britannica, 1952.

Rahner, Hugo. *Greek Myths and Christian Mystery*. Translated by Brian Battershaw. New York: Harper & Row, 1963.

Ranlet, Philip. "The British, the Indians, and Smallpox: What Actually Happened at Fort Pitt in 1763?" *Pennsylvania History: A Journal of Mid-Atlantic Studies* 67 (2000) 427–41.

Reppert, Victor. *C. S. Lewis's Dangerous Idea: In Defense of the Argument from Reason*. Downers Grove, IL: InterVarsity, 2003.

Rescher, Nicholas. *Free Will: An Extensive Bibliography*. With the collaboration of Estelle Burns. Frankfurt: Ontos, 2010.

———. *A Useful Inheritance: Evolutionary Aspects of the Theory of Knowledge*. New York: Rowman & Littlefield, 1990.

Reynolds, L. D., and N. G. Wilson. *Scribes and Scholars: A Guide to the Transmission of Greek and Latin Literature*. 2nd ed. Oxford: Clarendon, 1974.

Riedel, Stefan. "Edward Jenner and the History of Smallpox and Vaccination." *Baylor University Medical Center Proceedings* 18 (2005) 21–25.

Ring, Merrill. *Beginning with the Pre-Socratics*. Palo Alto, CA: Mayfield, 1987.

Roesler, Christian. "Are Archetypes Transmitted More by Culture than Biology? Questions Arising from Conceptualizations of the Archetype." *Journal of Analytical Psychology* 57 (2012) 223–46.

Rogers, Katherin. "St. Anselm of Canterbury on God and Morality." *Monist* 105 (2022) 309–20.

Ronan, Colin A. "The Origins of the Reflecting Telescope." *Journal of the British Astronomical Association* 101 (1991) 335–42.

Rosenberg, Alex. *How History Gets Things Wrong: The Neuroscience of Our Addiction to Stories*. Cambridge, MA: MIT Press, 2018.

Rosenstand, Nina. *The Moral of the Story: An Introduction to Ethics*. 7th ed. New York: McGraw Hill, 2013.

Ross, Hugh. *A Matter of Days: Resolving a Creation Controversy*. Colorado Springs: NavPress, 2004.

Rowbotham, Samuel (Parallax). *Zetetic Astronomy: Earth Not a Globe! An Experimental Inquiry into the True Figure of the Earth*. Bath: Haywood, 1865.

Rudolph, Kurt. "Religionsgeschichtliche Schule." Translated by Matthew J. O'Connell. In *The Encyclopedia of Religion*, edited by Mircea Eliade, 12:293–96. New York: Macmillan, 1987. https://www.encyclopedia.com/environment/encyclopedias-almanacs-transcripts-and-maps/religionsgeschichtliche-schule.

Russell, Jeffrey Burton. *Inventing the Flat Earth: Columbus and Modern Historians*. New York: Praeger, 1991.

Rutherford, David. "Lactantius Philosophicus? Reading, Misreading, and Exploiting Lactantius from Antiquity to the Early Renaissance." In *Essays in Renaissance Thought and Letters: In Honor of John Monfasani*, edited by Alison Frazier and Patrick Nold, 475–89. Brill's Studies in Intellectual History 241. Leiden: Brill, 2015.

Sacks, Oliver. *The Man Who Mistook His Wife for a Hat and Other Clinical Tales*. New York: Simon & Schuster, 1985.

Sagan, Carl. *Contact*. New York: Simon and Schuster, 1985.

———. *The Demon-Haunted World: Science as a Candle in the Dark*. New York: Random House, 1995.

———. *Pale Blue Dot: A Vision of the Human Future in Space*. New York: Ballantine, 1997.

Sagan, Carl, and Ann Druyan. *Comet*. New York: Random House, 1985.

Sampson, Philip. *Six Modern Myths About Christianity and Western Civilization*. Downers Grove, IL: InterVarsity, 2001.

Schadewald, Robert J. "Looking for Lighthouses." *Creation/Evolution* 12 (1992) 1–4. https://dsimanek.vialattea.net/litehous.htm.

Schindler, Samuel. *Theoretical Virtues in Science: Uncovering Reality Through Theory*. Cambridge: Cambridge University Press, 2018.

Seife, Charles. *Zero: The Biography of a Dangerous Idea*. New York: Penguin, 2000.

Severian. *Homilies on Creation and Fall*. Translated by Robert C. Hill. In *Ancient Christian Texts: Commentaries on Genesis 1–3: Severian of Gabala and Bede the Venerable*, 23–94. Downers Grove, IL: InterVarsity, 2010.

Severin, Tim. *The Brendan Voyage: A Leather Boat Tracks the Discovery of America by the Irish Sailor Saints*. New York: McGraw-Hill, 1982.

Shakespeare. William. *The Life of King Henry the Fifth*. Edited by Alfred Harbage. In *The Complete Pelican Shakespeare*, 741–77. New York: Viking, 1977.

Shklovskii, Iosef S., and Carl Sagan. *Intelligent Life in the Universe*. San Francisco: Holden-Day, 1966.

Siemens, David F., Jr. "Misquoting Tertullian to Anathematize Christianity." *Philosophia Christi* 5 (2003) 563–66.

Simon, Pierre. *Le contrôle des naissances*. Paris: Payot, 1966.

Slagle, James R. *Artificial Intelligence: The Heuristic Programming Approach*. New York: McGraw-Hill, 1971.

Slagle, Jim. *The Epistemological Skyhook: Determinism, Naturalism, and Self-Defeat*. London: Routledge, 2016.

———. *The Evolutionary Argument Against Naturalism: Context, Exposition, and Repercussions*. London: Bloomsbury Academic, 2021.

———. "Indicators and Depictors." *Philosophical Forum* 48 (2017) 91–107.

———. "Yes, Eliminative Materialism Is Self-Defeating." *Philosophical Investigations* 43 (2020) 199–213.

Smith, Jonathan Z. "Dying and Rising Gods." In *The Encyclopedia of Religion*, edited by Mircea Eliade, 4:521–27. New York: Macmillan, 1987. https://www.encyclopedia.com/environment/encyclopedias-almanacs-transcripts-and-maps/dying-and-rising-gods.

Smith, Mark S. "The Death of 'Dying and Rising Gods' in the Biblical World: An Update, with Special Reference to Baal in the Baal Cycle." *Scandinavian Journal of the Old Testament* 12 (1998) 257–313.

Sober, Elliot. "Evolution Without Naturalism." In *Oxford Studies in Philosophy of Religion*, edited by Jonathan L. Kvanvig, 3:187–221. New York: Oxford University Press, 2011.

Sowell, Thomas. *Intellectuals and Society*. Rev. ed. New York: Basic, 2011.

Squires, Roger. "Solipsism." In *The Oxford Companion to Philosophy*, edited by Ted Honderich, 838–39. New York: Oxford University Press, 1995.

Stark, Rodney. *For the Glory of God: How Monotheism Led to Reformations, Science, Witch-Hunts, and the End of Slavery.* Princeton, NJ: Princeton University Press, 2003.

Stein, John. "Bartolomeo Platina." New Advent, 1911. From *The Catholic Encyclopedia*, vol. 12. http://www.newadvent.org/cathen/12158a.htm.

Stevens, Wesley M. "The Figure of the Earth in Isidore's 'De Natura Rerum.'" *Isis* 72 (1980) 268–77.

Sungenis, Robert. *Galileo Was Wrong, the Church Was Right.* 3 vols. 11th ed. State Line, PA: CAI, 2015.

——. *Not by Faith Alone: A Biblical Study of the Catholic Doctrine of Justification.* Goleta, CA: Queenship, 1997.

Swinburne, Richard. *The Existence of God.* Rev. ed. New York: Oxford University Press, 2004.

Tennemann, Wilhelm Gottlieb. *Geschichte der Philosophie.* Vol. 8, pt. 2. Leipzig: Barth, 1811.

Tertullian. *On the Flesh of Christ.* Translated by Dr. Holmes. In *Ante-Nicene Fathers*, edited by Alexander Roberts and James Donaldson, 3:521–43. 1885. Repr., Peabody, MA: Hendrickson, 1995.

——. *The Prescription Against Heretics.* Translated by Peter Holmes. In *Ante-Nicene Fathers*, edited by Alexander Roberts and James Donaldson, 3:243–67. 1885. Repr., Peabody, MA: Hendrickson, 1995.

Thadani, Krishan M. "The Myth of a Catholic Religious Objection to Autopsy: The Misinterpretation of *De Sepulturis* During the Renaissance." *National Catholic Bioethics Quarterly* 12 (2012) 37–42.

Thijssen, Hans. "Condemnation of 1277." *Stanford Encyclopedia of Philosophy*, Winter 2018. Edited by Edward N. Zalta. https://plato.stanford.edu/archives/win2018/entries/condemnation/.

Thorndike, Lynn. *History of Magic and Experimental Science.* 8 vols. New York: Columbia University Press, 1923–58.

Tornau, Christian. "Saint Augustine." *Stanford Encyclopedia to Philosophy*, Summer 2020. Edited by Edward N. Zalta. https://plato.stanford.edu/archives/sum2020/entries/augustine/.

Trigg, Roger. *Beyond Matter: Why Science Needs Metaphysics.* West Conshohocken, PA: Templeton, 2015.

Turner, William. "St. Vergilius of Salzburg." New Advent, 1912. From *The Catholic Encyclopedia*, vol. 15. http://www.newadvent.org/cathen/15353d.htm.

Vainio, Olli-Pekka. *Cosmology in Theological Perspective: Understanding Our Place in the Universe.* Grand Rapids: Baker Academic, 2018.

Van Helden, Albert. "The Invention of the Telescope." *Transactions of the American Philosophical Society*, n.s., 67 (1977) 5–67.

——. *Measuring the Universe: Cosmic Dimensions from Aristarchus to Halley.* Chicago: University of Chicago Press, 1986.

Van Woudenberg, René. "Self-Defeat, Inconsistency, and the Debunking of Science." *Social Epistemology Review and Reply Collective* 9 (2020) 83–92. https://social-epistemology.com/2020/06/26/self-defeat-inconsistency-and-the-debunking-of-science-rene-van-woudenberg/.

Wagner, Richard. *Richard Wagner's Letters to August Roeckel.* Translated by Eleanor C. Sellar. Bristol: Arrowsmith, 1897.

Ward, Peter D., and Donald Brownlee. *Rare Earth: Why Complex Life Is Uncommon in the Universe*. New York: Copernicus, 2000.

Warfield, B. B. *Counterfeit Miracles*. New York: Scribner's Sons, 1918.

Watterson, Bill. *Homicidal Psycho Jungle Cat*. Calvin and Hobbes Collection 10. Kansas City: Andrews and McMeel, 1994.

———. *There's Treasure Everywhere*. Calvin and Hobbes Collection 11. Kansas City: Andrews and McMeel, 1996.

Watts, Edward J. *City and School in Late Antique Athens and Alexandria*. Berkeley: University of California Press, 2006.

Wessel, Susan. *Cyril of Alexandria and the Nestorian Controversy*. Oxford Early Christian Studies. New York: Oxford University Press, 2004.

Whitcomb, John C., and Henry M. Morris. *The Genesis Flood: The Biblical Record and Its Scientific Implications*. Phillipsburg, NJ: P&R, 1961.

White, Andrew Dickson. *A History of the Warfare of Science with Theology in Christendom*. 2 vols. London: Macmillan, 1895.

White, Lynn. "The Historical Roots of Our Ecological Crisis." *Science* 155 (1967) 1203–7.

Whitehead, Alfred North. *Science and the Modern World*. New York: Macmillan, 1925.

Wilberforce, Samuel. Review of *On the Origin of Species, by Means of Natural Selection; or the Preservation of Favoured Races in the Struggle for Life*, by Charles Darwin. *Quarterly Review* 108 (1860) 225–64.

Wildberg, Christian. "Neoplatonism." *Stanford Encyclopedia of Philosophy*, Winter 2021. Edited by Edward N. Zalta. https://plato.stanford.edu/archives/win2021/entries/neoplatonism/.

Willard, Dallas. *The Divine Conspiracy: Rediscovering Our Hidden Life in God*. San Francisco: HarperCollins, 1998.

Williams, George C. *Plan and Purpose in Nature: The Limits of Darwinian Evolution*. London: Phoenix, 1996.

Witherington, Ben, III. "Birth of Jesus." In *Dictionary of Jesus and the Gospels*, edited by Joel B. Green et al., 60–74. Downers Grove, IL: InterVarsity, 1992.

———. *The Jesus Quest: The Third Search for the Jew from Nazareth*. 2nd ed. Downers Grove, IL: InterVarsity, 1997.

Wolfe, Tom. *Hooking Up*. New York: Farrar, Straus, and Giroux, 2000.

Wright, N. T. *Jesus and the Victory of God*. Minneapolis: Fortress, 1997.

Wurmbrand, Richard. *Tortured for Christ*. Bartlesville, OK: Living Sacrifice, 1967.

Yamauchi, Edwin. *Pre-Christian Gnosticism: A Survey of the Proposed Evidences*. Grand Rapids: Eerdmans, 1973.

Yates, Frances A. *Giordano Bruno and the Hermetic Tradition*. Chicago: University of Chicago Press, 1964.

Zirkle, Conway. "Natural Selection Before the 'Origin of Species.'" *Proceedings of the American Philosophical Society* 84 (1941) 71–123.

Index

217 propositions, 187
2001: A Space Odyssey (film), 113–14
abacus, 12–13
abiogenesis. See origin of life
Abraham, 155
Abrahamic religions, xiii, 43n17, 152, 158–59, 163, 165, 221. See also Christianity; Islam; Judaism; Judeo-Christianity
absolutely nothing
    vacuum, 12, 37, 56, 163, 195
    zero, 11–12, 64–65
ACLU, 122–24
ad hocness, 43, 138, 205–8, 233
ad hominem fallacy, 82, 246n31
Adam and Eve, 62, 76–77, 132, 145–47, 170–71, 173, 235
    as historical/nonhistorical, 15, 145–47, 235
Adonis, 239
aether. See quintessence
afterlife, 210–11, 253n50
age of the universe/earth, 123–24, 144–45. See also eternity; time
agnosticism, 117, 119, 122
Albertus Magnus, 131, 162, 187
alchemy, 11, 15, 70, 213
Alexandria, 74, 89–93
all other things being equal, 139, 195–96, 202, 209
allegory of the cave, 88
Also Spracht Zarathustra (symphonic poem), 114

Amalricians, 186
America, Americans, 77, 126, 192, 250
    native, 4–5, 126
Anaximander, 130–31
ancient Christianity's views on science, 179–81
ancient Near East, 25, 236, 245, 249n44
anesthesia, 71, 76–78
animals, 50, 72, 79, 114n2, 126, 130–31, 133, 147, 153, 157–59, 162, 164, 170, 180, 249
    as souls, 158
    authority over, 40, 62
    cruelty toward, 62–63
    death, origin of, 147. See also death before the Fall
    human beings as. See human beings as rational animals
    rights of, 63
    value of. See value, animals
Animaniacs, 54
Anselm, 181, 217n9
Anthropic Principle, anthropic coincidences, 60–61, 63–65, 144. See also fine tuning
anthropocentrism, 43, 45, 49, 165, 229
    anthropo-peripheralism, 45, 165
    See also geocentrism vs. egocentrism
Antioch school, 27–28, 30
antipodes, antipodeans, 21–23, 27, 35, 172–73, 177, 192
    spiritual status of, 21–23, 172–73

antiquity (ancient western civilization), 7, 10, 12, 19–20, 30–31, 39, 55, 74, 181–82, 195
   late, 28, 30
   See also Christianity, ancient; Greek empire; Roman empire
anti-Semitism, 246
apologia, xi, xiii
appealing to authority, xii
Aquinas, Thomas, 41, 62, 101, 131, 153–54, 158, 162, 186–87, 217n9, 218–19
Arabic language, 29, 183–84
Arabic numerals, 12–13
archetypes, 247
Aristarchus of Samos, 39
Aristotle/Aristotelianism, xn2, 3, 18, 37–38, 55, 101, 105, 117, 128–32, 157–58, 181, 184, 186–87, 232
   condemnations of, 186–87. See also 217 propositions
   cosmology, 10, 18, 30, 37–38, 41–42, 46, 61, 101, 104–5
   physics, 44, 99, 128–30, 162
   view on animals, 62
artificial experiment, 188
artificial intelligence, 136, 174–75
Association for Biblical Astronomy, 51
astrology, 10–11, 70, 98
   Christian condemnation of, 10–11, 98
astronomy, 11, 25, 33–34, 39, 46, 55, 58–59, 101
atheism, 89–90, 114, 117, 138, 143, 157–59
Athena, 237
Athens, 84–87, 181
   gods of, 85, 87
atomic physics, 24–25, 248–49. See also Rutherford-Bohr model
atomism, 162
Attis, 239
Augustine, 35, 89, 117, 131, 153, 158, 173, 180, 186, 192, 217n9
aut Deus aut homo malus, 253
authority, xii, 39–40, 62, 106, 180, 208, 252, 254

   of the Catholic church, 17, 40–41, 110
   of the clergy, 17, 31, 41, 102
   See also appealing to authority
autopsies, 75. See also dissection, human
Averroes, Averroism. See Ibn Rushd; Latin Averroists

Babylon, 155
Bacon, Francis, 7–8, 188
Bacon, Kevin, *passim*
Bacon, Roger, 98, 100, 131
baptism, 23, 75, 236, 238–39, 248
Barberini, Maffeo. See popes, Urban VIII
Barbour, Ian, xv
Barth, Karl, 189n40
Basil, 181–82
Bayrakdar, Mehmet, 131
beauty, 113–14, 206, 232–33
Bedford canal, 33–34
Beeblebrox, Zaphod, 54n2
belief (content and causal properties of), 143, 223–24
Bellarmine, Robert, 103–5
Bergson, Henri. See *élan vital*
bias, 6, 14–15, 228, 236. See also confirmation bias
Bible, xiv, 10, 22–26, 40, 42, 47–48, 52, 58, 62, 79, 110, 123, 137–38, 147–48, 154, 158, 166, 168n22, 169, 189, 192, 202, 209, 212, 222, 235–36
Bible and science. See science and the Bible
biblical cosmology, 43–44, 68
biblical interpretation, xiv, 27–29, 41, 48n26, 81, 102–3, 110–11, 144–45, 192–93. See also Antioch school; literalism; phenomenal language
biblical worldview. See Judeo-Christianity, worldview
big bang cosmology, 60, 99
Big Fish in a Small Pond Myth, 54–56, 150–51, 153

Index

biology, 61, 113, 116, 122, 125, 132, 141, 249
birth control. See contraception
birthday cake analogy, 129–30
Black Death, 1–2, 184
black hole, 61, 99. See also white hole
blind faith, 182, 225, 230
    Christian condemnation of, 182, 230
Blount, Elizabeth, 33–34
Boethius, 19, 39, 55
    Consolation of Philosophy, 19
Bohmian mechanics, 200n16
Bonaventure, 131, 162
Boniface (the saint not the pope), 23
Boorstin, Daniel, 35–36
boundary issues, 110, 124
Brahe, Tycho, 39–40, 42, 93, 101. See also geocentrism, Tycho's modified
Brendan the Navigator, 250
Brodie, Benjamin, 119
Browne, Thomas, 227
Bruno, Giordano, 40, 42, 58–59, 93–97, 101, 103, 162, 189
    imprisonment and execution, 96–97, 101
    On the Infinite Universe and Worlds, 95
Bryan, William Jennings, 121–25
Buber, Martin, 59
Bultmann, Rudolph, 246
Bunny, Bugs, 35
Bunyan, John, 233
Buridan, John, 44
Burnett, James, 132
Byzantine empire, 183

Caesar, Julius, 92
Caesar, Theodosius, 92–93
Cairns-Smith, A.G., 136
calendar day interpretation, 144–45
Calcidius, 49n30
Calvin, John; Calvinism, 42, 60n16, 74, 244n25
Campanella, Tommaso, 104
Campbell, Joseph, 238, 240n13, 241–43
Camus, Albert, 79
capitalism, 115
Carolingian Renaissance, 183
Catholicism/Catholic church, xn2, 9, 41–42, 50–51, 71, 75, 82, 93–96, 99, 101–2, 105, 107, 109–11, 185, 190
causality, xiii, 128–31
    efficient, 128–29, 130n8, 131, 159
    material, 129–31, 159
    formal, 128–30
    final. See final causality
    instrumental, 130, 137
    paradigmatic/exemplary, 130, 159
center, literal vs. metaphorical, 43, 45, 50, 150. See also Myth of Dethronement
ceteris paribus. See all other things being equal
centrism. See geocentrism; heliocentrism; infernocentrism
Chalmers, Robert, 135
champagne and tuxedo analogy, 221–22
Chardin, Teilhard de, 134
charms, 242–43
Chaucer, Geoffrey, 57
chemistry, 11, 61, 141
childbirth, 76–77, 213
chloroform. See anesthesia
Christ Myth, 235–55
Christian Science, 81
Christian theology/theologians, xiv, 12, 41–42, 50–51, 56, 62, 89–90, 115, 171–72, 179–82, 185–89, 213–15, 244n25, 245–46
    and science. See science and Christian theology
    as queen of the sciences, 180. See also handmaiden doctrine
    medieval, 8, 188. See also Aquinas; Augustine; science and medieval theology
Christian/church tradition, 10, 17, 80, 147, 151, 165, 170, 172, 186, 189, 216, 230, 243
Christianity, Christians
    ancient, 30–31, 45, 55–56, 58, 60, 69, 131, 145, 173, 179–82, 227
    evangelical, 77, 117, 124–25

Christianity, Christians (*cont.*)
  medieval, 10–11, 23, 30–31, 55–58, 60, 69, 73, 131, 173, 181, 183–88, 214–15, 218–19
  unanswered questions in, 22–23, 170, 172.
    See also Catholicism; Eastern church; Orthodoxy; Protestantism; Nestorianism; Western church
Christianity and Islam, 11, 29, 35, 72–73, 183–85, 219–22
Christocentrism, 45
Chrysostom, John, 28, 30
Church Fathers. See Christianity, ancient
church hierarchy in Catholicism, 31, 46, 83, 102, 108, 185
Churchland, Patricia, 71, 76, 78, 81–82
Cicero, 66
civil laws, 196
clerics, 108, 110, 185
coherentism, 6n7
colonialism, 115
Columbus, Christopher, 17, 31, 35, 50, 83, 234, 248n39
Comet Myth, 8–10, 13–14, 16, 249
Condé, Maryse, 137
confirmation bias, 6, 32
Conflict Thesis/Myth, xi, 190–91, 193
conspiracy theories, 159, 206, 242, 247–48, 252
Contact (book and film), 65–66
contraception, 71, 82
contrived explanation, 43, 138, 146, 200, 205, 208, 232, 241–43. See also ad hocness
controlling nature, 94, 188–89, 209–10
Copernicus, Nicolaus/Copernicanism, 29, 40, 42–43, 46, 49, 58–59, 93, 101, 103, 110, 116, 142, 161
  On the Revolutions of the Heavenly Spheres, 42, 103
corruption, incorruption (of the universe), 44, 46, 99
Cosmas Indicopleustes, 28–30, 35
cosmic inflation, 60

cosmology. See Aristotle, cosmology; Bible, cosmology; modern cosmology; Ptolemy, cosmology
cosmos. See universe
Council of Ephesus, 27
Counter-Reformation, 103
courts, ecclesiastical vs. civil, 185
COVID, 81
cowpox. See smallpox and cowpox
cow-tipping, 2, 227
Craig, William Lane, 255n54
crap, 35, 85, 158
  holy crap, 14, 86
creation (i.e. what was created), 58, 62, 99, 153–54, 160, 169, 171. See also universe
creation event, 99, 147. See also days of creation; young-earth creationism
creation myths, 235–36
credo quia absurdum, 182
Cremonini, Cesar, xn2
Crossan, John Dominic, 255n54
culture wars, 121
curse of the Fall. See Fall of humankind, curse of
Cuvier, Georges, 134
Cyril (Patriarch of Alexandria), 91

d'Alembert, Jean le Rond, 190
Danielson, Dennis, 161, 177
Dante, 19, 45, 57
  Divine Comedy, 19
Dark Ages, 7, 89, 93, 187, 213
Darrow, Clarence, 121–23
Darwin, Charles, 24, 33, 113, 115–19, 121, 130, 132, 135, 141, 144, 157, 190
  On the Origin of the Species, 113, 116, 119, 190
  religious beliefs of, 117–18
  time and culture of, 115–16, 118, 144, 248
Darwin, Erasmus, 132, 135
Darwinism, 39, 115–25, 131–32, 135–48, 190. See also evolution
  evidence for, 116–17, 159
  social. See social Darwinism

Index

DaVinci, Leonardo, 76n24
day-age theory, 123, 144–45
days of creation, 122–23, 144–45
Dayton Tennessee, 121–24. See also Scopes Monkey Trial
De Sepulturis, 72–73
death before the Fall, 147–48, 170–71
decentralizing the earth, 49, 149, 157, 159. See also Myth of Dethronement
deferents. See spheres, planetary
Dennett, Daniel, 60n17, 142
Descartes, René, 111
determinism, 11, 200–201
    physical/mechanistic, 63, 200
    See also atomism; Newton, Newtonianism
developmentalism, 113, 132
Diderot, Denis, 132–33
Digges, Leonard, 100
Digges, Thomas, 40, 58–59, 93–94, 95n40, 100
Dilman, İlham, 86n7
Diodore of Tarsus, 27–28, 30
Dionysius the Areopagite. See pseudo-Dionysius
disease. See prayer for disease; smallpox; vaccination
dissection, human, 72–76, 98. See also human body
divine hiddenness, x
domination thesis, 62–63
dominion thesis, 62–63
Draper, John William, 120–21, 191
Druyan, Anne, 8, 10
Duhem, Pierre, 39
Dutch Reformers, 189n40
dying and rising gods. See Christ Myth

Earth
    not in motion, 44, 47–48, 103, 105–6, 111, 150, 160. See also geocentrism; motion
    revolution of, 93
    rotation of, 44
    spatial insignificance of, 55–56, 58, 66–67, 155–56. See also Big Fish in a Small Pond Myth

Eastern church, 28–29
economics, 12, 87, 143, 206–7
Eden, 57, 145, 150, 160, 170, 235. See also Adam and Eve
Edwards, Jonathan, 79
egocentrism. See geocentrism vs. egocentrism
Egypt, 18, 26, 95–96, 137, 222, 238–39. See also Alexandria
Einstein, Albert, 232
élan vital, 134–35
elements
    four terrestrial, 37, 44, 71, 180
    fifth. See quintessence
eliminative materialism, 71
Empedocles, 131
empirical testing/proof, x, 209–10, 247
empiricism, 8
Enlightenment, 7, 63, 188, 190, 233–34
environmentalism, 63. See also dominion thesis; value, universe's
Epicureans, 162
epicycles, 38
eppur si muove, 100, 110
Eratosthanes, 18
Eriugena, John Scotus, 90
eschatology, 98
eternity, 58
eucharist. See Last Supper
Eudoxus, 18
eugenics, 115–16, 123, 125–27. See also social Darwinism
Europe, Europeans, 4–5, 11–12, 17, 29, 46, 75, 77, 120, 125–26, 150, 160, 185–88, 198, 212–15, 246
Euthyphro Dilemma, 216–17, 220
evangelism/evangelicals, 22, 77, 117, 124–25, 179
Everitt, Nicholas, 63
evidence, ix-x, 7, 10, 42, 52, 61, 101, 105, 116–17, 138, 140, 145–46, 159, 176, 182, 185, 203–9, 227, 229–30, 237, 242–43, 248
    inconclusive, 44, 101
    no evidence that, 4, 9, 12, 42, 107, 110, 176, 237
evil. See problem of evil

Evil Twin Theory, 200, 207, 255. See also Jesus, resurrection of
evolution, 15, 39, 112–48, 157–59, 163–64, 174, 190, 206, 223–24, 248–49
   and the Bible, 144–46
   as theory of everything/grand unified theory, 113, 125, 248
   beauty of, 113–15
   evidence of, 116–17, 159, 206
   plus, 138, 143, 158–59
   pre-Darwinian, 130–34. See also Lamarckism
   See also Darwinism
ex cathedra, 106, 111
excommunication, 23, 32, 187
   of Halley's comet. See Comet Myth
execution, 54, 72, 74, 86–87, 93, 96, 99, 101, 198–200, 208, 254
Exodus, 222
explanatory theory, hypothesis, 139–40, 255
explanatory virtues, 43, 116, 202–8. See also ad hocness; probability/inprobability
extraterrestrial life, 66, 162–64
extraterrestrial intelligence, 3n2, 152–53, 160, 161–77, 192
   religions of, 164–65
   spiritual status of, 160, 165–73, 176
extremophiles, 163

faith
   alone. See sola fide
   and reason, 181, 193, 214–15, 220, 229–31
   and science, 193, 212, 229–30. See also science and the Bible; science's dependence on religion
   blind. See blind faith
Fales, Evan, 249n44
Fall of humanity, 21–22, 147, 167. See also death before the Fall; original sin
   cosmic fall, 22, 169–72, 175–76
   curse of, 76–77, 170
Ferngren, Gary, 70

fictionalism, 48–49, 142–43. See also saving the appearances
fideism. See blind faith
final causality, 129–31, 133–36, 138–43. See also teleological argument
   in biology, 141–43. See also Watchmaker argument
fine structure constant, 63–65
fine tuning, 60–61, 63–65, 144
first humans. See Adam and Eve; origin of humanity
Flat Earth Myth, 31–37, 52, 248n39
flat earth, xv, 17–36
   and the Bible, 23–26
   Christian acceptance of, 26–35, 52
   societies, 33, 50
Flatliners (films), 210–11
flood myths, 235–36
forms, Platonic, 88
Foscarini, Paolo, 103
Francis of Meyronnes, 44
Frazer, James George, 246–48
free thought, 89, 97
free will, xiii, 11, 147, 179n3, 200–201, 204. See also determinism
Freud, Sigmund, 160
Frogface. See Socrates
functions, biological, 131, 141–43. See also final causality in biology
fundamentalism, 124, 127
Fundamentals of the Christian Faith, 124–25

Galactic Bethlehem Model, 160, 168, 172, 176
Galen, 71, 74
Galilean moons, 99–101. See also Jupiter
Galilei, Galileo, x, 40–43, 49, 98–111, 149, 234
   Dialogue Concerning the Two Chief World Systems, 104–6
   imprisonment of, 106–7
   interrogation of, 105–6
   torture of, 99, 106, 108–10
   trials of, 31, 102–6, 111
Galileo affair, 41–42, 98–111
gap theory, 123, 144–45

gaps arguments. See God of the gaps
general revelation, 189, 192–93
geocentrism, 37–52, 101, 104–5, 149–50, 156, 162, 162, 233
    and the Bible, 47–48, 102
    evidence for, 42–43, 101–2, 105
    galactocentrism, 51–52
    societies, 50–52
    Tycho's modified, 40, 42, 93, 101
    vs. egocentrism, 40, 43, 45, 49–50, 52, 149
    See also anthropocentrism; Galilei, Galileo; heliocentrism; infernocentrism; Myth of Dethronement
germ theory, 70, 232
Gilbert, William, 40
giraffe's neck as example, 133–34. See also Lamarck, Lamarckism
Gliddon, George R., 15
globes, 20–21
globularism, 34. See also round earth
globus cruciger, 19–20
Gnostic Redeemer Myth, 246
Gnosticism, 236, 246
God
    as ground of morality, 151, 217, 220, 226
    as ground of rationality, 151, 153, 217–18, 220, 222–26
    as ground of reality/existence, 114, 151, 153, 217, 226, 253
    as ground of value, 151, 153
    as love, 155, 231
    as metaphysically necessary, 140–41, 220, 225–26
    existence/nonexistence of, ix-x, 45, 59, 61, 63–65, 68, 89, 119, 134, 138, 140–41, 144, 146, 157, 179, 189, 205, 207, 220n14, 224–26, 231. See also agnosticism; atheism
    freedom of, 197, 204, 211, 218, 220
    goodness of. See Euthyphro Dilemma
    greatness of, 57–58, 66
    intellect vs. will, 218–19
    loved by, 46, 62–63, 114, 151–52, 155–58, 176, 218, 231
    nature/properties of, 140–41, 179, 189, 220n14, 224–26
    omnipotence of, ix, 66, 146, 197, 219–20
    omnipresence of, 43n17
    omniscience of, ix, 211
    transcendence of, 220–21, 225
    See also Prime Mover
God of the gaps, 206–8
    Darwinism of the gaps, 207n28
    Naturalism of the gaps, 207–8
    See also miracles, ad hocness of
God or Church not pleased by blood, 72–73, 220
good guys and bad guys, 83, 103, 119, 120–21
Gould, Stephen Jay, 14–15
Graeco-Roman biography, 251
Grand Duchess Christina, 102
grand unified theory, evolution as. See evolution as theory of everything
Grant, Robert Edmond, 134
gratuitousness, 63, 138–40. See also ad hocness
grave robbing, 75–76
gravity, 60–61, 195–96
    Newtonian vs. Aristotelian, 99, 232. See also Aristotle, physics; Newton, Newtonianism
Greek empire, 18
Greek language, 29, 119, 183
Greek philosophy and science, 12, 62, 178–79, 181–82, 215
Greek writings, 29, 183–84
Green, J.R., 120
gunpowder, 100, 234

Haddon, A.C., 242
Haeckel, Ernst, 249
Halley's Comet. See Comet Myth
handmaiden doctrine, 179–80
Hannam, James, xv, 70, 76n24, 227–28
Harriot, Thomas, 40, 100
Hawking, Stephen, 63–65, 68, 98–99, 226

heart, function of, 141, 143. See also functions, biological
heaven, 25, 46, 57, 68, 79, 157n12, 180, 210, 246
heavens, the. See motion, heavens; space; spheres, celestial; universe
heavy metal, 224
Hebrew language, 158n15, 166n16
Hebrews. See Israelites
Hegel, G.W.F., 100
heliocentrism, 39–43, 46–49, 93–95, 99, 101–6, 109–11, 157, 232. See also anthropocentrism; Galilei, Galileo; geocentrism; infernocentrism; Myth of Dethronement
hell, 45–46, 114, 117, 149–50, 210, 233
Henswell, J.S., 119
heresy, 17, 27, 29n29, 35, 91n23, 96, 105–6, 108, 111, 186
Hermeticism, 94–95, 246
Hero's Journey, 240
heterodoxy, 91, 104
Hipparchus of Nicaea, 38
Hippocrates, 71
Historical Jesus, 244, 248, 252–55
  third quest for, 244, 246
Historical Question, 213–16
historical explanation, 204–5
historical writing, 250–52
History of the Conflict between Religion and Science. See Draper, John William
History of the Warfare of Science with Theology in Christendom. See White, Andrew Dickson
Hitchens, Christopher, 114
Hodge, Charles, 117
Holy Spirit, 95
Hooker, Joseph, 119–20
Hope, Elizabeth, 118
horror vacui, 59
horse teeth, 6–8
Horus, 239
hospitals, 80–81, 209–10, 227
human beings
  as moral agents. See moral agency
  as rational agents, 153, 217, 222–25
  as rational animals, 117, 153, 157–59, 173, 205n24
  autonomy of, 6, 80
  created in the image of God. See image of God
  death of, 147–48. See also death before the Fall
  evolution of, 113–14, 123–25, 130, 145–46, 157–59, 223–24. See also social Darwinism
  geographical extent of. See antipodes, antipodeans
  significance/insignificance of. See image of God; Myth of Mortification
human body/bodies, 72, 78, 80, 141, 188
  dead, 72–76, 98, 239. See also dissection, human
Humanae Vitae, 82
Hume, David, 130n8, 143, 198–202, 225. See also miracles, Hume's argument against
humoral theory, 70–71, 232
Hunter, George William, 122, 125–26
Huxley, Aldous, 190, 253n50
Huxley, Julian, 190
Huxley, T.H., 118–21, 125, 157, 190
Huxley-Wilberforce debate, 118–21
Hypatia of Alexandria, 89–93
hyperbole, 25, 181–82
hypostatic union, 253

Ibn Rushd, 184, 186–87
ideology, 87, 136, 142
idolatry, 220–21
illusion, 57, 81, 101, 143
image of God (imago Dei), 22, 72–73, 152–60, 166–68, 170–76, 217, 221
  interpretations of, 153–54
immunization. See vaccination
imprisonment, 41, 86, 93, 96, 98–99, 101, 106–7, 109, 121–22
incarnation, Jesus'. See Jesus, incarnation of
Index of Prohibited Books, 103, 106, 111
inductive inference, 119, 139, 198–99, 225–26
  problem of, 198–99, 225

Index

infernocentrism, 45–46, 150
  diabolocentrism, 45
  See also anthropocentrism; Galilei, Galileo; geocentrism vs. egocentrism; heliocentrism; Myth of Dethronement
Inherit the Wind, 122. See also Scopes monkey trial
inoculation, 79. See also vaccination
Inquisition, 96, 99, 103, 105–10
  inquisitors, 17, 31, 109
  palace, 106–7
Institute for Creation Research, 51
intellectual advancement/development, 120, 183–85, 188
intellectual history, ix, 184
intellectual influence, 73
intellectuals, xi-xii, 173, 230–32, 235. See also monstrous races
Internet, 2, 20, 35
Iron Age, 242–43
Irving, Washington, 31
Isidore of Seville, 30, 35
Isis (goddess), 238–39
Islam, Muslims, xiii-xiv, 11–12, 29, 35, 50, 81, 94, 131–32, 137, 148, 152, 165, 174, 183–85, 213, 219–22
Islamic theology, 25, 72–73, 216, 219–23, 224–25
Israel, 23–24, 26, 245, 253
Israelites, 26, 48, 137, 222
  as chosen people, 155
Italy, 73, 96

Jāḥiẓ, 131–32
Jaki, Stanley, 212
Janssens, Louis, 80
javelin argument. See Lucretius
Jeannière, Abel, 80
Jenner, Edward, 79
Jerusalem, 181, 200, 207, 253, 255
Jesuits, 99, 190
Jesus, 22, 68, 87, 92, 147, 150, 154, 160, 168, 171–73, 178–79, 194, 198, 209, 213, 217, 235–37, 240–41, 243–49, 252–55
  appearances of, 204, 255
  ascension of, 25, 46, 68–69
  atonement, 22, 46, 167–69, 171–73
  burial of, 198–99, 207, 255
  crucifixion and death of, 171, 198–200, 207–10, 236, 240–41, 245, 254
  divinity of, 95, 253–54. See also hypostatic union
  empty tomb of, 198–99, 204, 207, 255
  incarnation of, 22, 95, 160, 167–69, 171–72, 176, 178–79, 182, 217, 253. See also logos, Jesus as
  resurrection of, 25, 46, 68, 194, 198–200, 204, 207–8, 235–37, 239, 241, 245, 248, 249n44, 252, 254–55
  return of, 25–26
  temptation of, 26
  virginal conception of, 238
  See also Christianity; historical Jesus; salvation
Judaism, 40, 62, 94–95, 137, 151–53, 165, 196, 198, 213, 215n7, 219, 224, 244–45, 253
  first century, 244–45
Judeo-Christianity, 15, 41, 73, 144, 147, 151, 159, 243
  worldview, xiii, 17, 156, 214
Jung, Carl, 247–48
Jupiter, 18
  moons. See Galilean moons
justification, doctrine of, 51
Juvenal, 198n7

Keefe, Donald, 80
Kennedy, John F., 253n50
Kepler, Johannes, 39–40, 93–94, 189
Khoury, Theodore, 219–20
Kim, Jaegwon, 223n21
Kingsley, Charles, 117
knowledge, 6n7, 228–29
  gaps in. See God of the gaps
  vs. opinion, 88, 90
Koyré, Alexandre, 58–59, 94
Kubrick, Stanley. See *2001: A Space Odyssey*

Lactantius, 27, 30
Laertius, Diogenes, 18
laity, laymen, 41, 82, 102, 105
Lamarck, Lamarckism, 132–36
Laplace, Pierre-Simon, 8–9
Last Supper, 236, 238
Latin Averroists, 184, 187. See also Ibn Rushd
Latin language, 27, 29, 37, 79, 105, 119, 184, 227–28
Law of Conservation of Energy, 196–97
laws of nature (*not* natural law), 113, 115, 137–38, 140, 187, 194–97, 218–19. See also miracles
leaning tower of Pisa, 99–100
Leclerc, George-Louis, 132
Leopold-Loeb case, 122
Lewis, C.S., 115, 166–67, 176n33, 233, 236, 241, 250–51, 253n50
The Pilgrim's Regress, 233–34
Library of Alexandria, 89, 91–93
life, 48, 60–62, 64–65, 68, 79n34, 113, 135–36, 151–52, 158, 177
extraterrestrial. See extraterrestrial life
origin of. See origin of life
simple, 163–64. See also extremophiles
light as indicating value, 150, 160
lighthouses, 33–34
Lincoln, Abraham, 240
Lindberg, David, 179, 181, 192
Lipperhey, Hans, 100
literalism, 27
literary analysis, 249–52
literary framework interpretation, 145
literary genres, 249–52
Liu, Cixin, 175–76
logarithmic scale, 67
logos, 178–79
Jesus as, 172, 178–79, 217; see also Jesus, incarnation of
long 18th century. See Enlightenment
Lovejoy, Arthur, 45, 162
Lucas, J.R., 128
Lucretius, 95
Luther, Martin, 42
Lyons, Jonathan, 35

Maestlin, Michael, 40, 93
magic, 94, 188–89, 209, 239, 243. See also science vs. magic
magnetic force, 195–96
Maillet, Benoît de, 132–33
Man from Earth, The (film), 247–48
Mandaeanism, 246
Mandeville's Travels, 19
Mann, Charles C., 13
Manuel II Paleologus, 219–20
Manx, Isle of Man, 241–42
maps, 20–21, 24
T-O maps, 20, 150, 160
Mars, 18
life on, 163
martyrs, 72, 83–97
mass density, 60–61
matter, 12, 24, 44, 51, 60–61, 131, 136, 152, 152n5, 184, 197
Maupertius, Pierre Louis, 132
Mayr, Ernst, 142
McCormick, Richard, 80
McGrath, Alister, 51
mechanism (physical determinism). See determinism, physical
medicine, 70–82, 188
folk, 70
history of, 70–71, 227–28, 232
medieval era. See Middle Ages
Mediterranean, 26, 73
Mercury, 18, 56
metaphor, 45, 191n46
meteors, 163, 201, 203–4, 208
Metz, Gautier de, 19
Middle Ages, 1, 7, 19–20, 26, 28, 30, 39, 73–75, 94, 150, 183–89
early, 30, 90, 183
Milky Way galaxy, 52, 68. See also geocentrism, galactocentrism
Milton, John, 57
mind, xiii, 59, 118, 143, 174, 213, 215, 217, 222–25. See also problem of other minds
mind and brain, 143, 211, 229
mind-body problem. See belief, causal properties of

miracles, xiii, 69, 137, 194–208, 235, 252, 254–55
  ad hocness of, 205–8
  as signs, 196–97, 203, 205, 208
  assessment of, 197–202
  Hume's argument against, 198–202, 225
  performance of, 197
  probability of, 202–5
mirrors, 100
misinformation, 32
misogyny, 4
Mithras, 92, 237–38
Mlodinow, Leonard, 226
Möbius Earth, 21n11
models
  mathematical, 18, 38, 48–49
  scientific, 24, 40, 49, 248
modernity (modern era), 7, 19, 40, 72, 74, 94, 128, 132, 184, 187–89, 250–51
modern cosmology, 58, 65n31. See also science, modern
monogenism, 15, 132
monstrous races, 173
Montesquieu, 132
Monty Python, 54, 189n39
  subtle references to, 19, 96, 133
Moon, 10, 18, 27, 40, 52, 53, 56, 63, 99–101, 145, 162, 180. See also sphere of the moon
Moon landings, 159, 201, 203
moons of the outer planets, 163. See also Galilean moons
møøse, majestic, 133
moral agency, 11, 22, 153, 155, 158, 166–67, 169–70, 217–18, 231
moral laws, 195, 218. See also value
Moreland, J.P., 223n21
Morris, Henry, 123
mortification, 149, 157–60, 176–77. See also Myth of Mortification
Morton, Samuel George, 14–15
Moses, 95, 137
motion, 130n8
  as indicating value, 18, 44, 150, 160
  circular, 18, 44
  downward/centerward, 43–46, 162

heaven's/sky's, 18, 38–40, 43–44, 47–48, 180
  upward, 46
  See also Earth not in motion
Muhammad, 221
Murray O'Hair, Madalyn, 13–14, 16, 82
Murray, William J., 13
music references, 113–14, 159, 224n22
Muslims. See Islam
mutations, natural vs. rational, 134–35, 138–39
my sister, xvi, 35, 133n19
mystery religions, 236–41, 244–45
myth(s), xii–xiii, 1, 6, 37, 71, 112, 173, 237n5. See also mythology; urban legend
Myth of Dethronement, 43–44, 49, 52, 149, 153, 161
Myth of Mortification, 149–51, 157–60, 175–77
mythology, 1, 235–51
  as literary genre, 249–52, 254
  Egyptian, 238–39
  Near East, 236–37, 245, 249n44
  Norse, 241–43
  world, 236, 238, 244, 249
mythos, 178–79

Nag Hammadi literature, 246
Nagel, Thomas, 207n28, 231, 248n41
Narnia, 176n33
natural history. See biology
natural law (*not* laws of nature), 78, 82, 113, 115n6
natural philosophy. See proto-science
natural processes, God's use of, 137–38
natural selection, 33, 131–32, 223
  vs. rational selection, 134–35
naturalism, 32, 116, 119, 142, 156–57, 159, 205, 207–8, 223–24
naturalist fallacy, 115n6
nature, 2, 10, 49–50, 54, 57, 62, 115, 115n6, 119, 142, 162, 171, 175, 196–97, 202, 205, 214, 223
  as a means to understand God, 118, 189. See also general revelation
  control over, 94, 114, 120, 188–89, 209

nature (cont.)
    understandability of, 214–15, 215n6, 217–18, 225–26
    See also creation; laws of nature; miracles, Hume's argument against; universe
Nazism, 125
near-death experiences, 210–11
negative theology, 90
Negev, 26
Neoplatonism, 89–90, 130, 181
Neptune, 19, 53
Nestorianism, 29
new Heaven and Earth, 147, 171
New Testament. See Bible
New Testament scholarship, 238, 244–46, 248, 251–55
Newman, John Henry, 139
Newton, Isaac, 99, 200, 232
Newtonianism, 138, 200
Nicholas of Cusa, 43n17, 58, 59, 94, 162
Nietzsche, Friedrich, 54, 86n8, 114
nihilism, 234
Nile River, 239
Nott, Josiah C., 15

Ockham's Razor, 139–40. See also William of Ockham
Old Testament. See Bible
Oracle at Delphi, 84
orbits, 18, 38, 40, 93, 99, 101, 180. See also spheres, planetary
Oresme, Nicole, 44, 162
Orestes (Prefect of Alexandria), 91
Origen, 89, 145–46, 162
origin of humanity, 123–24, 145–46, 158, 193. See also Adam and Eve
origin of life, 135–36, 144, 164, 173, 207
original sin, 21–22, 52, 76–77, 147, 166–75
    transmission and application of, 167–69, 171–72, 174–75
    See also fall of humanity
Orthodoxy, Christian, 71, 82, 203
Osiris, 238–39
other minds. See problem of other minds
overprovision, 142

paganism, 192, 238, 240, 245–46
pain, 76–78. See also problem of evil
    Congenital Insensitivity to (CIP), 78
Pannenberg, Wolfhart, 172, 208
pantheism, 90, 95–97
parables, 145, 179, 235–36. See also stories
Park, Katherine, 74
Parmenides, 12, 18
Parousia. See Jesus, return of
parting of the Red Sea, 137
Pascal, Blaise, 59, 155–56
Paul (apostle), 90, 154
Penrose, Roger, 64–65
persecution, 72, 74, 83, 87, 98, 179
phenomenal language, 24–25, 47
Philipse, Herman, 191n46
Philolaus, 18
Philoponus, John, 29
Philosophical Question, 213–26
physical laws. See laws of nature
physical mechanisms, 138–39
    vs. method, 132–33, 135
physics, 12, 24–25, 48–49, 122, 138, 141. See also Aristotle, physics; Newton; quantum physics
physiology, 120, 142
Planck length, 67
planets, 18–19, 38–40, 53, 61, 63, 93, 150, 162–63. See also spheres, planetary
Plantinga, Alvin, 146–47, 218–19, 223n21
Platina, Bartolomeo, 8–9
Plato, Platonism, 17–18, 88–90, 104, 129, 162, 181–82, 186–87, 216. See also Socrates
Platonic forms. See forms, Platonic
player piano analogy, 140
Plotinus, 89. See also Neoplatonism
plurality of worlds, 93–96, 162, 187
Polkinghorne, John, xv
polygenism, 15
popes
    Benedict XVI, 219–20
    Boniface VIII (the pope not the saint), 72–73
    Calixtus III, 8–9, 16, 249

Gregory IX, 186
Gregory XVI, 106
John XXI, 187
John Paul II, 51n32, 99
Leo XII, 79–80
Paul V, 101
Sylvester II, 11
Urban VIII, xn2, 41, 101, 104–5, 110
Zachary, 23
porcupines, 2, 227
postmodernism, 188
prayer, 9, 222, 239
   efficacy of, 209–11
   for disease, 80–81
premodern art and literature, 57
premodern cosmology. See Ptolemaic cosmology
premodern physics. See Aristotle, physics
premodern science. See proto-science
prescription vs. description, 195
pre-Socratics, 37, 71, 88, 130–31, 162
Prime Mover, 38, 43. See also Aristotelian cosmology; God
prime reality, 58, 177. See also God
principle of plenitude, 162, 176
probability, improbability, 64–65, 68, 199–205
   predictive sense vs. explanatory sense, 203–5
problem of evil, x, 78, 117, 146–48
   and evolution, 146–48
   origin of evil, 147–48
problem of other minds, 165–66, 229
Proctor, Richard, 33–34
prolegomena, xi, 174
propaganda, 32, 81, 187, 191
Protestantism, 41, 50–51, 72, 82, 96, 102, 110, 122, 124, 190, 202–3, 246. See also Reformation
proto-science, xiv, 41, 70, 88–89, 181, 213
Pseudo-Dionysius, 90
psychoanalysis, 160, 247
Ptolemaic cosmology, 10, 18–19, 38–39, 41–43, 47, 53–56, 61, 63–65, 68–69, 93–94, 101–2, 104
Ptolemy, 10, 12, 35, 38–39, 53, 55, 64
Pythagoras, Pythagoreanism, 18, 181

quantum physics, 200–201
quills. See porcupines
quintessence, 37, 39, 44, 56, 59, 61, 101
Qur'an, 81, 183, 221

racism, 14–15, 120–21, 127. See also eugenics; social Darwinism
Rare Earth Hypothesis, 163–64
rationalism, 8
rationes seminales, 131
reason and rationality, 3, 66–67, 83, 112, 114, 151–54, 165–67, 174, 178–80, 182, 193, 213–15, 217–21, 229–30
   faith and, 229–30. See also fideism
   science and, x, 193, 229
   pure vs. practical, 219–20
   See also extraterrestrial intelligence; God as ground of rationality; image of God; human beings as rational agents
rebirth. See Renaissance; resurrection
Recorde, Robert, 100
recommended reading, xn1, 3n2, 21n11, 21n12, 39n5, 43n16, 54n3, 86n7, 123n32, 145n44, 146n46, 154n9, 165n12, 168n20, 175n30, 178n2, 188n36, 212n1, 213n4, 217n9, 223n21, 229n3, 230n8, 253n50, 255n54
Red Sea, 137
Reformation, 41, 96, 102, 110, 202–3. See also Catholicism; Counter-Reformation; Protestantism
Reitzenstein, Richard, 246
Religio Medici, 227
religion (general), x–xiv, 8, 10, 13–16, 32, 35, 41, 59, 65n31, 80–82, 87, 89, 97, 112, 114, 119, 121, 123, 125, 135, 143, 149, 155, 161, 16–65, 168n23, 174, 177–79, 194, 210, 213–15, 221, 233–34, 236, 247
   and science. See science and religion
   See also Abrahamic religions; Christianity; Islam; Judaism; Judeo-Christianity

religionsgeschichtliche Schule, 237, 245–46, 248
religious particularism, 95, 155, 164–65
Renaissance, 7, 27, 58, 73–74, 94, 98, 187–88
Reppert, Victor, 63n26
Rescher, Nicholas, 134
resurrection, 180, 236–41, 246–48, 251. See also Jesus, resurrection of
revelation, 22, 119, 154, 164, 166, 169–70, 189, 192–93, 202–3, 231. See also general revelation
Rheticus, Georg Joachim, 40, 42, 46, 49, 93
Ring around the Rosy, 1–2
Ring of the Niebelung, 113. See also Wagner, Richard
ritual, 244
Rockel, August, 113
rogue waves, 201, 203, 208
Roman, 20, 80, 92–93
Roman empire, 12, 19, 26, 92–93, 131, 183, 236
    fall of, 28, 89, 93, 183
Roman inquisition, 96, 108, 110
Roman numerals, 11–13
Roman philosophy and science, 10, 55, 62, 181–83. See also Greek philosophy and science
Rome, 26, 104, 106–8
Rothmann, Christopher, 40
round Earth, 17–37, 53, 55, 234
    evidences of, 18–19
    circumference of, 18
    See also globularism
Rowbotham, Samuel ("Parallax"), 32–34
rule of thumb, 4
Rutherford-Bohr model, 24, 248–49

Saadia Gaon, 50
Sacrabosco, Johannes, 19
Sagan, Carl, 8, 10, 40–41, 46, 63–65, 68, 149, 151n4. See also *Contact*
Saint-Hilaire, Étienne Geoffrey, 134
salvation, 22–23, 46, 50–51, 150, 160, 166–69, 173, 175, 180, 236. See also antipodes, spiritual status of; extraterrestrial intelligence, spiritual status of
Satan, 7, 26, 45, 147–48, 171. See also infernocentrism, diabolocentrism
Saturn, 18, 53
saving the appearances, 48, 103, 142. See also fictionalism
Scandinavia, 241–42
science fiction references, 3n2, 21n11, 39n5, 43n16, 47, 65, 78, 165n12, 168, 175–76, 175n30, 229n3
science
    advance of, 31–32, 61, 71, 82, 116, 119, 132, 142, 183–84
    and medieval theology, 187, 214–16, 218–20
    and the Bible, 40, 76, 106, 111, 138, 144–46, 212
    and religion (exact phrase), x–xiii, 5–7, 13, 31–32, 52, 71, 76, 82, 99, 123, 145, 149, 178, 190–91, 193, 210, 212, 221, 228, 232
    as ideology, 136, 142. See also scientism
    contemporary, xiv, 24–25, 55, 70, 124, 128–29, 234
    glory of, 15
    life, 141. See also biology
    modern, 130, 141, 161, 188, 214–15, 219. See also modernity
    of the day, 10, 37–39, 41, 179, 182
    patterns of, 83, 113, 135–36, 157, 248
    presuppositions of, 213, 217, 229
    professional/nonprofessional, 78, 119
    premodern. See proto-science
    secular, 41, 181–82. See also secularism
    social, 141
    societies, 116, 119
    Western, 74–76, 219
    value of, 55, 179–81
    vs. magic, 94–95, 188–89
science's dependence on religion, xiii, 212–26
scientific analysis, 139
scientific argument, 18, 67, 105, 120

scientific discovery/discoveries, x, 32–33, 53–55, 65n31, 76–77, 89, 98–101, 124–25, 149–51, 157, 163, 177, 190, 193, 200, 204–5, 228–29
scientific evidence, x-xi, 42, 44, 101, 105, 116–17, 138, 145, 159, 176, 204, 206–7, 229–30
scientific ideas, 13–14, 97, 132, 163, 207n28
Scientific Revolution, 7, 72, 89, 94, 162, 187–89, 212–13, 215, 219
scientism, xi, 136, 142, 193, 228–29
scientists' fear of other scientists, 42, 93, 101, 116
Scientology, 81
Scopes monkey trial, 121–27
Scopes, John, 121–24
Scripture. See Bible
sculpture analogy, 129
sea turtles, 142
secularism, xiv, 116, 119
selection process, 134–35, 139, 141, 223–24. See also natural selection
Seneca, 49
senses, bodily, 3
Serapeum, 92–93
serial killers, 157
Servetus, Michael, 72, 74
Seventh Day Adventism, 122
Severian of Gabala, 28, 30
Severin, Tim, 250
Shai-Hulud, 47
sharia law, 72–73
Sherlock Holmes, 199
shooting stars, 101. See also meteors
siege of Belgrade, 8–9
Siena, 107
Simocatta, Theophylactus, 29
Simon, Pierre, 80
Simplicius of Cilicia, 105
Simpson, James Young, 76–77
sin, 44, 46, 76, 171, 240. See also original sin
skepticism, 12, 14, 16, 65, 249
    selective, 82. See also confirmation bias
skulls, 14–15
Slagle, Jim (the other one), 175
slavery, 26, 63, 120
    abolition of, 119
smallpox, 4–5, 79–80
    and cowpox, 79
    as divine judgment, 79
    infected blankets, 4–5
    See also inoculation; vaccination
Smith, Jonathan Z., 240–41, 247n37
social Darwinism, 115–16, 119–20, 125–27
    and evolution, 125
social sciences, 141, 246
Socrates, xi, 17–18, 84–89, 97, 216
    execution of, 86–87
    trial of, 86
    See also Plato, Platonism
sola fide, 50–51. See also faith; Reformation
sola Scriptura, 102
solar wind, 163
solipsism, 229
South England Legendary, 19, 39, 69
space, 47, 56, 60–61
    awful waste of, 66, 162
    empty, 24, 37, 39, 56, 59, 61, 68, 152. See also absolutely nothing, vacuum
    finite/infinite, 56, 58–60, 99. See also unbounded universe
Spain, 26, 183
Spartans, 84
spheres, celestial, 18–19, 37–38
    concentric, 18, 30, 37, 53, 56
    planetary, 18–19, 38
    of the fixed stars, 18, 43, 53, 55–57, 69, 94
    of the moon, 18–19, 37, 44, 56–57
Spong, John Shelby, 68–69
spontaneous generation, 135–36. See also origin of life
Spurgeon, C.H., 117
Squishington, Bartholomew, 84–85
Stanton, Graham, 251n47
star of Bethlehem, 10
Stark, Rodney, 212–13

stars, 10–11, 18, 27, 38–40, 46, 49, 52, 60–61, 93, 145, 150, 180
    fixed, 38, 55. See also sphere of the fixed stars
    size of, 53, 55–56, 61, 180
    wandering. See planets
    See also spheres
static universe, 201
Stevin, Simon, 40, 93
Stoics/Stoicism, 18, 131, 181
stories, 1, 4–6, 8–9, 11, 14, 16, 41, 72, 82–83, 89–91, 93, 100, 118, 121–22, 137, 144–47, 155, 178–79, 190, 235–38, 240–41, 244, 247–52. See also myth; parables; urban legend
Strauss, Richard. See *Also Spracht Zarathustra*
straw man arguments, 68, 161
sublime, 67
Suez Canal, 137
Sun, 18, 27–28, 38–40, 43, 46–47, 53–54, 63, 88, 93, 95, 99, 145, 149, 180
    stopped in sky, 48
    See also heliocentrism
Sungenis, Robert, 51
sunspots, 99, 101
supernaturalism, 12, 25, 32, 48n26, 68–69, 80, 135–36, 143, 158, 164, 174, 194, 196, 204–5, 207, 236, 238, 252. See also miracles; naturalism
supreme being. See God.
surgery, 73–74. See also medicine
syncretism/anti-syncretism, 244–45

tabernacle universe, 25, 28–29. See also flat earth
Tatian, 181–82
taxonomy, 131–32
technology, 26, 36, 113, 174–75, 183, 188
    early medieval, 183
teleological arguments, 144. See also Watchmaker Argument
teleology. See final causality
telescopes, xn2, 18, 33–34, 99–100

Tempier, Étienne, 187
Tennemann, Wilhelm, 100
Tennessee, 121–22, 124–25
tent, universe as, 25, 27–28, 47. See also tabernacle universe
Tertullian, 181–82, 230
Thales, 178–79, 225
theocentrism, 45, 165
theodicies, 146–48
Theodore of Mopsuestia, 27–28, 30
theology. See Christian theology; God; Islamic theology
Theophilus (Bishop of Alexandria), 92
third quest. See historical Jesus, third quest for
Thomism. See Aquinas, Thomas
Thor (not the superhero), 241–43
thorns and thistles, 77, 170
tide, 101
time, 51–52, 56, 60, 69, 99, 116, 131, 144
Tolosani, Giovanni, 46
Total Perspective Vortex, 53–54
Tracy, Spencer, 122
transformation of species. See mutations
translation movement, 29
Trinity, 52, 253
Trismegistus, Hermes. See Hermeticism
Tristram, Henry Baker, 120
Tuscany, 107, 110
Typhon, 239

Übermensch, 114
unbounded universe, 58, 93–96. See also space, finite/infinite
unimaginable sizes/distances, 53–56, 58–60, 150
United Kingdom, 33, 81, 118, 242
United States, 118, 121–22
universe
    age of, 123–24, 145. See also days of creation
    as tabernacle. See tabernacle universe
    created for God, not humanity, 62, 68. See also theocentrism
    perfection/imperfection of, 10, 46, 99, 101. See also quintessence

size of, 39, 52, 53–69, 150–51, 156–57, 159, 176–77. See also Big Fish in a Small Pond Myth
universities, 185–87, 213, 227–28
   Harvard, 13
   of Bologna, 73
   of Chicago, 36
   of Göttingen, 245
   of Padua, 73
   of Paris, 186–87
   of Toulouse, 186
   Oxford, 100, 119
Uranus, 19, 53
urban legend, 1–8, 12, 14–15, 17, 31–32, 40–42, 71, 93, 98, 112, 149, 191, 235
utilitarianism, 151

vaccination, 78–81
vacuum. See absolutely nothing
value, 66–68, 149–61
   animals', 62–63, 152, 158
   humanity's, 60, 152–61, 176
   intrinsic/extrinsic, 62–63, 179–81
   objective, 151, 153, 156
   sources of, 151–53
   spectrum, 155–56, 160
   universe's, 152.
   vs. fact, 62–63
   See also dominion thesis; image of God; light as indicating value; motion as indicating value; Myth of Mortification
Vatican, 73, 99
vegetation cycle, 240–41
Venerable Bede, 19
Venice, 96
Venus, 18, 56
   phases of, 99–100
Vergilius of Salzburg, 23
Vesalius, Andreas, 72, 74, 98
Victorian society, 115
violence, 4, 89, 93, 220
virgin birth, 236–38, 248, 251. See also Jesus, virginal conception of
visions, 24, 26, 48n26, 122, 233, 250
vivisection, 74

Voltaire, 190

Wagner, Richard, 113, 115n6
wait for no man. See time; tide
Wallace, Alfred Russel, 33
Warfield, B.B., 117
Watchmaker argument, 144
Weismann, August, 134
Western church, 29, 96, 184
Western civilization, 17–18, 29–30, 38, 113, 117, 119, 125, 183, 189
what has Athens to do with Jerusalem?, 181–82
Whitcomb, John, 123
white hole, 51–52
white supremacy, 14–15, 120, 125, 127
White, Andrew Dickson, 81, 191
White, Ellen, 122
Whitehead, Alfred North, 214–15
Wilberforce, Samuel, 118–21. See also Huxley-Wilberforce debate
Wilberforce, William, 119–20
William of Ockham, 218–19, 222, 224–25. See also Ockham's Razor
Williams, George C., 142
witchcraft, 94, 189. See also magic
Wolfe, Gene, 78
worldview(s), 1, 5, 156, 164, 168n23, 193, 212–13, 215–16, 223, 226, 228, 231
   biblical/Judeo-Christian. See Judeo-Christianity, worldview
   science as. See scientism

xenomorphs. See extraterrestrial life

young-earth creationism, 51, 58, 122. See also calendar day interpretation

Zeitgeist (film), 248
zero. See absolutely nothing
Zero Myth, 11–14
zetetic astronomy, 33. See also Rowbotham, Samuel
Zeus, 237
Zúñiga, Diego de, 40, 42

www.ingramcontent.com/pod-product-compliance
Lightning Source LLC
Chambersburg PA
CBHW032052220426
43664CB00008B/964